QUANTIZED
ALGEBRA
AND PHYSICS

NANKAI SERIES IN PURE, APPLIED MATHEMATICS AND THEORETICAL PHYSICS

Editors: S. S. Chern, C. N. Yang, M. L. Ge, Y. Long

Published:

Proceedings of the International Workshop on
Quantized Algebra and Physics

Tianjin, China   23 – 26 July 2009

Nankai
Series in
Pure,
Applied
Mathematics
and
Theoretical
Physics

Vol. 8

QUANTIZED ALGEBRA AND PHYSICS

Edited by

Mo-Lin Ge
Nankai University, China

Chengming Bai
Nankai University, China

Naihuan Jing
North Carolina State University, USA &
South China University of Technology, China

World Scientific

NEW JERSEY · LONDON · SINGAPORE · BEIJING · SHANGHAI · HONG KONG · TAIPEI · CHENNAI

Published by

World Scientific Publishing Co. Pte. Ltd.
5 Toh Tuck Link, Singapore 596224
USA office: 27 Warren Street, Suite 401-402, Hackensack, NJ 07601
UK office: 57 Shelton Street, Covent Garden, London WC2H 9HE

British Library Cataloguing-in-Publication Data
A catalogue record for this book is available from the British Library.

Nankai Series in Pure, Applied Mathematics and Theoretical Physics — Vol. 8
QUANTIZED ALGEBRA AND PHYSICS
Proceedings of International Workshop

ISBN-13 978-981-4340-44-1
ISBN-10 981-4340-44-8

Printed in Singapore by World Scientific Printers.

PREFACE

Algebra and physics have long been intertwined. During the last century, starting from H. Weyl's treatment on quantum mechanics, Yang-Mills theory, Yang-Baxter equations, and topological quantum field theory etc, mutual developments have reached an unprecedented stage that almost all major developments in mathematics have drawn critical ideas from physics.

The era of quantum groups further demonstrated the mutual propulsion between mathematics and physics. When Drinfeld and Jimbo introduced the concept of quantum groups, Kac-Moody algebras provided one of the key motivations. In FRT formulation, the theory of Lie groups and Yang-Baxter equation, which can be viewed as a braid relation, played the key role. Perhaps one indicator of the deep relationship between algebra and quantum theory is that quantum algebra has been one of the most popular areas on arXiv, each month hundreds of papers have been posted and they mirrored the early popularity of hepth in physics arXiv in a large scale.

During July 23–26, 2009 the international workshop on quantized algebra and physics was held at Chern Institute of Mathematics at Nankai University in Tianjin, China. The organizing committee consists of Mo-Lin Ge (Nankai), Chengming Bai (Nankai) and Naihuan Jing (NCSU and SCUT). About 50 experts attended the activities from four continents: Asia, Europe, North America and Australia. Parallel sessions were run in the four mornings and two afternoons with a common talk. Roughly speaking the programs are divided into the following areas:

1) Quantum groups and in particular Yangians

2) Quantum mechanics and quantum field theory

3) Kac-Moody algebras and in particular affine Lie algebras

4) Combinatorial aspects of representation theory

5) Algebraic groups

6) Other important classes of algebras, including Hall algebras, Virasoro-Schrödinger algebras, Briman-Murakami-Wenzl (BMW) algebras etc.

The workshop has also provided a rich environment of discussion between Chinese graduate students and their American counterpart. Seven graduate students from North Carolina State University and more than thirty Chinese graduate students and young teachers from Nankai, Chinese Academy of Sciences, Peking University, East China Normal University, South China University of Technology Northeastern Normal University, Xiamen University, Tianjin University, Dalian Institute of Technology, Changshu Institute of Technology, Jiangsu University, University of Science and Technology of China, Communication University of China, Civilaviation University of China, Northeastern University at Qinhuangdao, and Shanghai Normal University have attended the activities.

This volume contains some partial record of talks and contributions at the workshop. We thank all the anonymous referees for their reports which have made this proceedings possible. The editors also thank Nankai Series in Pure, Applied Mathematics and Theoretical Physics to include this volume. We also thank World Scientific for their patience for the reviewing process.

The workshop is graciously supported by the staff members at Chern Institute of Mathematics. The organizers are grateful to financial support of Chern Institute of Mathematics, Nankai University, the Ministry of Education of China, the Ministry of Science and Technology of China and National Natural Science Foundation of China. N. Jing would also like to thank US National Science Foundation and National Natural Science Foundation of China for partial support.

Mo-Lin Ge, Tianjin
Chengming Bai, Tianjin
Naihuan Jing, Guangzhou

CONTENTS

Program At A Glance

	July 22	July 23 Section A	July 23 Section B	July 24 Section A	July 24 Section B	July 25 Section A	July 25 Section B	July 26 Section A	July 26 Section B	July 27
Chair		*N. Jing*	*D. Meng*	*Y. Su*	*F. Li*	*Y. Zhu*	*H. Yamada*	J. Du	C. Dong	
8:30-9:20		M. Ge		C. Dong	S. Hu	B. Bakalov	L. Li	K. Zhao	H. Yamada	
9:30-10:20		H. Rui	K. Misra	A. Molev	L. Ji	Y. Gao	F. Li	R. Zhang	H. Zhang	Leave
10:30-11:00	Regis-tration	Break								
11:00-11:50		Y. Su	L. Guo	J. Beier	Y. Zhu	Y. Chen	M. Rosso	C. Sun	P. Peng	
11:50-14:30		Lunch+Rest								
Chair		*A. Molev*	*M. Rosso*	*K. Zhao*	*X. Xu*	Rest		Rest		
14:30-15:20		S. Tan	N. Hu	J. Du	A. Helminck					
15:30-16:20		G. Helminck	X. Xu	D. Nakano	J. Hu					
16:30-17:00		Break								
17:00-17:50		A. Savage	B. Zhang		X. Chen					

● Time arrangement in the morning on **July 23**:
8:30-8:40 **Opening Remarks**   8:40-9:30 M. Ge   9:40-10:10 **Photo and Break**
10:10-11:00 **Section A: H. Rui**   **Section B: K. Misra**
11:10-12:00 **Section A: Y. Su**   **Section B: L. Guo**

Supported by

- Chern Institute of Mathematics, Nankai University
- Ministry of Education, China
- National Natural Science Foundation of China (10621101)
- NKBRPC (2006CB805905), Ministry of Science and Technology, China
- SRFDP (200800550015), Ministry of Education, China

July 22, Wednesday

8:30-22:00 **Registration** at Jiayuan Hotel, Nankai University

July 23, Thursday

Chair: Naihuan Jing (North Carolina State University, USA)
8:30-8:40 **Opening Remarks** (Lecture Hall 216, Shiing-Shen Building)

Yiming Long, Director, Chern Institute of Mathematics, Tianjin
8:40-9:30 Mo-Lin Ge (Nankai University, China)

Yang-Baxter equation and topological quantum field theory
9:30-10:10 Photo and Break Time
Section A: (Lecture Hall 216, Shiing-Shen Building)

Chair: Naihuan Jing (North Carolina State University, USA)
10:10-11:00 Hebing Rui (East China Normal University, China)

Blocks of Birman-Murakami-Wenzl algebras
11:10-12:00 Yucai Su (University of Science and Technology of China)

A combinatorial identity associated with a new quantization of the Virasoro algebra
12:00-14:30 Lunch

Chair: Alexandre Molev (University of Sydney, Australia)
14:30-15:20 Shaobin Tan (Xiamen University, China)

Representations of Virasoro-Schrödinger algebras
15:30-16:20 Gerardus Helminck (University of Amsterdam, the Netherlands)

The geometry of integrable hierarchies of infinite matrices
16:30-17:00 Break Time
17:00-17:50 Alistair Savage (University of Ottawa, Canada)

Moduli spaces of sheaves and the boson-fermion correspondence

Section B: (Lecture Hall 214, Shiing-Shen Building)
Chair: Daoji Meng (Nankai University, China)

10:10-11:00 Kailash Misra (North Carolina State University, USA)

Imaginary Verma modules and Kashiwara algebras for $U_q(sl(2)^{\wedge})$
11:10-12:00 Li Guo (Rutgers University, Newark, USA)

O-operators and associative Yang-Baxter equations
12:00-14:30 Lunch

Chair: Marc Rosso (University Paris Diderot- Paris 7, France)
14:30-15:20 Naihong Hu (East China Normal University, China)

Modular quantization of Cartan type Lie algebras
15:30-16:20 Xiaoping Xu (Chinese Academy of Science, China)

Representation of Lie algebras and coding theory
16:30-17:00 Break Time
17:00-17:50 Bin Zhang (Sichuan University, China)

Renormalization on toric varieties

July 24, Friday

Section A: (Lecture Hall 216, Shiing-Shen Building)
Chair: Yucai Su (University of Science and Technology of China)
8:30-9:20 Chongying Dong (University of California, Santa Cruz, USA)

Parafermion vertex operator algebras
9:30-10:20 Alexandre Molev (University of Sydney, Australia)

Higher Sugawara operators for gl(n)
10:30-11:00 Break Time
11:00-11:50 Julie Beier (Mercer University, USA)

Demazure crystals of $U_q(\widehat{sl}(n))$
12:00-14:30 Lunch

Chair: Kaiming Zhao (Wilfrid Laurier University, Canada)
14:30-15:20 Jie Du (University of New South Wales, Australia)

A double Hall algebra approach to affine quantum Schur-Weyl theory
15:30-16:20 Daniel Nakano (University of Georgia, USA)

Cohomology of quantum groups

Section B: (Lecture Hall 214, Shiing-Shen Building)
Chair: Fang Li (Zhejiang University, China)
8:30-9:20 Sen Hu (University of Science and Technology of China)

Constructing an $N = 2$ superconformal vertex algebra with a generalized Calabi-Yau target

9:30-10:20 Lizhen Ji (University of Michigan, USA)

Coarse Schottky problem and equivariant cell decomposition of Teichmuller spaces

10:30-11:00 Break Time

11:00-11:50 Yongchang Zhu (Hong Kong University of Science and Technology, China)

Siegel-Weil formula for loop groups

12:00-14:30 Lunch

Chair: Xiaoping Xu (Chinese Academy of Science, China)

14:30-15:20 Alqysius Helminck (North Carolina State University, USA)

Orbit decompositions in symmetric spaces

15:30-16:20 Jun Hu (University of Sydney, Australia & Beijing Institute of Technology, China)

On a generalization of the Dipper-James-Murphy conjecture

16:30-17:00 Break Time

17:00-17:50 Xueqing Chen (University of Wisconsin, Whitewater, USA)

Root vectors, PBW and canonical bases of Ringel-Hall algebras and quantum groups

July 25, Saturday

Section A: (Lecture Hall 216, Shiing-Shen Building)

Chair: Yongchang Zhu (Hong Kong University of Science and Technology, China)

8:30-9:20 Bojko Bakalov (North Carolina State University, USA)

W-constraints for simple singularities

9:30-10:20 Yun Gao (York University, Canada & University of Science and Technology of China)

Lie triple systems and the Steinberg unitary Lie algebras

10:30-11:00 Break Time

11:00-11:50 Yuqun Chen (South China Normal University, China)

Gröbner-Shirshov bases and embedding of algebras

12:00-night Rest

Section B: (Lecture Hall 214, Shiing-Shen Building)

Chair: Hiro-Fumi Yamada (Okayama University, Japan)

8:30-9:20 Libin Li (Yangzhou University, China)

Introduction to Kaplansky's sixth conjecture

9:30-10:20 Fang Li (Zhejiang University, China)

Approach to Artinian algebras via natural quivers and generalized modulation theory
10:30-11:00 Break Time
11:00-11:50 Marc Rosso (University Paris Diderot-Paris 7, France)
Construction of irreducible representations of quantum groups and their characters from quantum shuffles
12:00-night Rest

July 26, Sunday

Section A: (Lecture Hall 216, Shiing-Shen Building)
Chair: Jie Du (University of New South Wales, Australia)
8:30-9:20 Kaiming Zhao (Wilfrid Laurier University, Canada)

Weight modules over the loop-Virasoro algebra
9:30-10:20 Ruibin Zhang (University of Sydney, Australia)

Quantum group actions on noncommutative algebras and invariant theory
10:30-11:00 Break Time
11:00-11:50 Changpu Sun (Institute of Theoretical Physics, China)
TBA
12:00-night Rest

Section B: (Lecture Hall 214, Shiing-Shen Building)
Chair: Chongying Dong (University of California, Santa Cruz, USA)
8:30-9:20 Hiro-Fumi Yamada (Okayama University, Japan)
Compound basis arising from the basic $A^{(1)}_{1}$-module
9:30-10:20 Hechun Zhang (Tsinghua University, China)
Composition series of tensor product and canonical basis
10:30-11:00 Break Time
11:00-11:50 Pan Peng (University of Arizona, USA)
Integrality structure and Chern-Simons/topological string duality
12:00-night Rest

July 27, Monday

Leaving Day

Thank you very much and welcome you to visit Chern Institute of Mathematics again!

A NOTE ON BRAUER-SCHUR FUNCTIONS

KAZUYA AOKAGE

Department of Mathematics, Okayama University,
Okayama 700-8530, Japan
Email: k.aokage@math.okayama-u.ac.jp

HIROSHI MIZUKAWA

Department of Mathematics, National Defence Academy of Japan,
Yokosuka 239-8686, Japan
Email: mzh@nda.ac.jp

and

HIRO-FUMI YAMADA

Department of Mathematics, Okayama University,
Okayama 700-8530, Japan
Email: yamada@math.okayama-u.ac.jp

1. Introduction

The aim of this note is to investigate certain symmetric functions which are related to modular representations of the symmetric group $\mathfrak{S}_n$.

Fixing an arbitrary prime number p, we define a family of polynomials $B_\lambda(t)$ indexed by p-regular partitions λ, and call them "Brauer-Schur functions". If we replace the variables $t = (t_1, t_2, \ldots)$ by the "original" variables $x = (x_1, x_2, \cdots)$ via

$$t_k = \frac{1}{k}(x_1^k + x_2^k + \cdots),$$

then we get a symmetric function $B^\lambda(x)$. Our problem is to examine the coefficients appearing in the expansion

$$B^\lambda(px) = \sum_\mu \kappa_{\lambda\mu} m_\mu$$

in terms of the monomial symmetric functions m_μ. It is conjectured that every $\kappa_{\lambda\mu}$ is an integer which is a multiple of $p^{\ell(\mu)}$. In this note we verify this phenomenon for the special case $\lambda = (n)$.

The third author thanks the organizers of Tianjin Conference for giving him a slot for presenting the paper.

2. Brauer-Schur functions

Let V denote the space of polynomials with infinitely many variables:

$$V = \mathbb{Q}[t_k; k \geq 1] = \bigoplus_{n=0}^{\infty} V_n,$$

where V_n is the subspace of homogeneous polynomials of degree n, with $\deg t_k = k$. In this note, p always denotes a fixed prime number. Put $V^{(p)} = \mathbb{Q}[t_k; k \geq 1, k \not\equiv 0 \,(\mathrm{mod}\,p)]$ and $V_{(p)} = \mathbb{Q}[t_{pk}; k \geq 1]$, so that $V = V^{(p)} \otimes V_{(p)}$. For a polynomial (or a formal power series) $F(t)$, define the p-reduction $F^{(p)}$ by killing all variables $t_{kp}, \ k = 1, 2, \ldots$. Namely

$$F^{(p)}(t) := F(t)|_{t_p = t_{2p} = \ldots = 0}.$$

It is well-known that the Schur functions form a basis for the space V. For a partition λ of n, the Schur function $S_\lambda(t)$ indexed by λ is defined by

$$S_\lambda(t) = \sum_{\rho} \chi_\rho^\lambda \frac{t_1^{m_1} t_2^{m_2} \cdots}{m_1! m_2! \ldots} \quad \in V_n.$$

Here the summation runs over all partitions $\rho = (1^{m_1} 2^{m_2} \cdots)$ of n, and the integer χ_ρ^λ is the irreducible character of λ of the symmetric group $\mathfrak{S}_n$, evaluated at the conjugacy class ρ. The "original" (symmetric) Schur function is recovered by putting

$$t_k = \frac{1}{k}(x_1^k + x_2^k + \cdots), \ k = 1, 2, \ldots.$$

Schur functions are ortho-normal with respect to the inner product

$$\langle F, G \rangle = F(\partial)G(t)|_{t=0},$$

where $\partial = (\frac{\partial}{\partial t_1}, \frac{1}{2}\frac{\partial}{\partial t_2}, \frac{1}{3}\frac{\partial}{\partial t_3}, \cdots)$.

A partition $\lambda = (\lambda_1, \ldots, \lambda_\ell)$ is said to be "p-regular" if there are no i's such that $\lambda_i = \ldots = \lambda_{i+p-1}$. The set of all p-regular partitions of n is denoted by $P^r(n)$. A partition $\lambda = (1^{m_1} \ldots n^{m_n})$ is said to be "p-class regular" if $m_{pk} = 0$ for any $k \geq 1$. The set of all p-class regular partitions of

n is denoted by $P^c(n)$. It is well-known that these two sets have the same cardinality. In fact, there is a natural bijection

$$G : P^r(n) \longrightarrow P^c(n)$$

defined as follows. Let $\lambda = (\lambda_1, \ldots, \lambda_\ell)$ be p-regular. If $\lambda_i = pk$, a positive multiple of p, then replace λ_i by $(k, \ldots, k)$, a p-repetition of k. Repeat this process to get a p-class regular partition $\widetilde{\lambda}$. For example, if $p = 2$ and $\lambda = (6, 4)$, then $\widetilde{\lambda} = (3, 3, 1, 1, 1, 1)$. It is easily observed that $\ell(\widetilde{\lambda}) - \ell(\lambda)$ is divisible by $p - 1$ for any $\lambda \in P^r(n)$. This map G is called the p-Glaisher map.

For a p-regular partition λ of n, the Brauer-Schur function $B_\lambda(t)$ indexed by λ is defined, similarly to the Schur function, by

$$B_\lambda(t) = \sum_\rho \phi_\rho^\lambda \frac{t_1^{m_1} t_2^{m_2} \cdots}{m_1! m_2! \cdots} \quad \in V_n^{(p)},$$

where the summation runs over all p-class regular partitions $\rho = (1^{m_1} 2^{m_2} \ldots)$ of n, and the integer ϕ_ρ^λ is the irreducible Brauer character of λ of the symmetric group $\mathfrak{S}_n$, evaluated at the p-regular conjugacy class ρ.

Again by putting $t_k = \frac{1}{k}(x_1^k + x_2^k + \cdots)$, we have a symmetric function $B^\lambda(x)$ of the "original" variables $x = (x_1, x_2, \ldots)$.

These Brauer-Schur functions are linearly independent, and hence form a basis for the space $V^{(p)}$. The space $V^{(p)}$ affords an irreducible highest weight representation of the affine Lie algebra of type $A_{p-1}^{(1)}$, and the Brauer-Schur functions are a weight basis ([1]). In this connection, one also sees that, when $p = 2$, $\{B_\lambda(D); \lambda^{(2)} \neq \emptyset\}$ gives a basis for the space of the Hirota polynomials for the BKP hierarchy ([6]), where $D = (D_1, \frac{1}{3}D_3, \frac{1}{5}D_5, \ldots)$ and $\lambda^{(2)}$ denotes the 2-core of the partition λ ([2]).

In cotrast with Schur functions, Brauer-Schur functions are not orthogonal with respect to the inner product $\langle \, , \, \rangle$. The dual basis is given by the "projective covers" of the Brauer-Schur functions, which we explain now.

For a partition λ of n, let $S_\lambda^{(p)}(t)$ be the p-reduced Schur function indexed by λ. Since these p-reduced Schur functions are elements of $V^{(p)}$, they can be expressed as linear combinations of Brauer-Schur functions:

$$S_\lambda^{(p)}(t) = \sum_{\mu \in P^r(n)} d_{\lambda\mu} B_\mu(t).$$

The coefficients are known to be non-negative integers. In fact $d_{\lambda\mu}$ is the number of occurrences of the modular irreducible module D_μ in the composition series of the Specht module indexed by λ, and is called the "decomposition number". Let $D_n = (d_{\lambda\mu})_{\lambda \in P(n), \mu \in Pr(n)}$ be the decomposition matrix, and consider the Cartan matrix $C_n := {}^t D_n D_n = (c_{\mu\nu})_{\mu,\nu \in Pr(n)}$. Let the projective cover of $B_\mu(t)$, for a p-regular partition μ, be defined by

$$\widetilde{B_\mu}(t) := \sum_{\nu \in Pr(n)} c_{\mu\nu} B_\nu(t).$$

Applying the representation theory of finite groups $(^4)$, one sees that

$$\widetilde{B_\mu}(t) = \sum_{\lambda \in P(n)} d_{\lambda\mu} S_\lambda(t).$$

Note that the right-hand side is a sum of (non-reduced) Schur functions. Again by the representation theory of finite groups $(^4)$, one has the orthogonality

$$\langle B_\lambda, \widetilde{B_\mu} \rangle = \delta_{\lambda\mu}.$$

Hence these are bases for the space $V^{(p)}$ dual to each other.

The product of two Schur functions is expressed as a linear combination of Schur functions:

$$S_\lambda S_\mu = \sum_\nu c_{\lambda\mu}^\nu S_\nu.$$

The structure constants $c_{\lambda\mu}^\nu$ are called the Littlewood-Richardson coefficients and known to appear in several different settings. For example, it is known that the branching of the restriction of the irreducible representation (λ, μ) of the hyperoctahedral group $W(B_n)$ to the subgroup $\mathfrak{S}_n$ is nothing but $c_{\lambda\mu}^\nu$. It is also known that the branching of the induced representation

$$\mathrm{Ind}_H^G \lambda \otimes \mu,$$

of the irreducible representation $\lambda \otimes \mu$ of $H = \mathfrak{S}_k \times \mathfrak{S}_{n-k}$ to $G = \mathfrak{S}_n$ is given by $c_{\lambda\mu}^\nu$. By the Frobenius reciprocity, this branching is nothing but the one of the irreducible representation ν of G restricted to H. One can expect similar results for the constants $\eta_{\lambda\mu}^\nu$ (resp. $\widetilde{\eta_{\lambda\mu}^\nu}$) defined by

$$B_\lambda B_\mu = \sum_\nu \eta_{\lambda\mu}^\nu B_\nu, \quad (\text{resp. } \widetilde{B_\lambda} \widetilde{B_\mu} = \sum_\nu \widetilde{\eta_{\lambda\mu}^\nu} \widetilde{B_\nu}).$$

This problem will be discussed in detail in our forthcoming paper.[5] Here we only remark the following.

Proposition 2.1. *Let p be a prime number, $G = \mathfrak{S}_n$ and $H = \mathfrak{S}_k \times \mathfrak{S}_{n-k}$. The number of occurrences of the modular irreducible representation ν of G in the composition series of the induced representation $\mathrm{Ind}_H^G \lambda \otimes \mu$ of a modular irreducible representation $\lambda \otimes \mu$, is given by $\eta_{\lambda\mu}^{\nu}$.*

Proof. It is an easy consequence of a general theory of characters. Let $\lambda \in P^r(k)$ and $\mu \in P^r(n-k)$. Let χ^λ (resp. χ^μ) be the irreducible character of λ (resp μ). If $\rho \in P^c(n)$ is a regular conjugacy class, then the induced character value is given by

$$\mathrm{ind}_H^G \chi^\lambda \otimes \chi^\mu(\rho) = z_\rho \sum_{\rho_1 \cup \rho_2 = \rho} z_{\rho_1}^{-1} z_{\rho_2}^{-1} \chi^\lambda(\rho_1) \chi^\mu(\rho_2),$$

where z_ρ denotes the order of the centralizer of an element having cycle type ρ. Denote the left-hand side by $\chi(\rho)$. Consider the "generating function"

$$\sum_\rho \chi(\rho) \frac{t_1^{m_1} t_2^{m_2} \cdots}{m_1! m_2! \cdots},$$

where the sumation runs over all class-regular partitions $\rho = (1^{m_1} 2^{m_2} \ldots)$. By a simple computation one sees that the right-hand side equals $B_\lambda(t) B_\mu(t)$. This proves the claim. $\qquad\square$

3. Cauchy type formula

Thanks to the duality of Brauer-Schur functions and their projective covers, we have the following "Cauchy formula" or, preferably, "Laplace formula". For a symmetric function $F(x)$, we write

$$\varphi(F(x)) := F(px),$$

where $px := (\underbrace{x_1, \ldots x_1}_{p}, \underbrace{x_2, \ldots x_2}_{p}, \ldots)$.

Theorem 3.1.

$$\sum_{\lambda \in P^r} \varphi(B^\lambda(x)) \widetilde{B^\lambda}(y) = \prod_{i,j} \frac{1 - x_i^p y_j^p}{(1 - x_i y_j)^p},$$

where the summation runs over all p-regular partitions.

Proof. Put

$$t_k = \frac{1}{k}(x_1^k + x_2^k + \cdots).$$

Consider

$$\eta(t, y_j) := \sum_{k=1}^{\infty} t_k y_j^k,$$

and its p-reduction $\eta^{(p)}(t, y_j)$. Then the desired formula is rewritten as

$$\sum_{\lambda \in P^r} B_\lambda(pt)\widetilde{B^\lambda}(y) = \exp\left(\sum_j \eta^{(p)}(t, y_j)\right),$$

where $pt = (pt_1, pt_2, \ldots)$. However the last equation is an easy consequence of the orthogonality (^{3}). $\qquad\qquad\qquad\qquad\qquad\qquad\qquad\qquad\qquad\square$

4. Monomial expansion

Since $\varphi(B_\lambda(x))$ is a symmetric function, it can be expanded in terms of the monomial symmetric functions

$$m_\mu(x) = \sum x^\alpha$$

summed over all distinct permutations of $\mu = (\mu_1, \ldots, \mu_\ell)$ (^{3}). For a p-regular partition λ and a partition μ, define the coefficients $\kappa_{\lambda\mu}$ by

$$\varphi(B_\lambda) = \sum_\mu \kappa_{\lambda\mu} m_\mu.$$

Set $K_n^{(p)} := (\kappa_{\lambda\mu})_{\lambda \in P^r(n), \mu \in P(n)}$ for a fixed prime p. Here we give $K_n^{(2)}$ and $K_n^{(3)}$ for some small n's.

$$K_3^{(2)} = \begin{array}{c|ccc} & (3) & (21) & (111) \\ \hline (3) & 2 & 4 & 8 \\ (21) & 2 & 8 & 16 \end{array}$$

$$K_4^{(2)} = \begin{array}{c|ccccc} & (4) & (31) & (22) & (211) & (1111) \\ \hline (4) & 2 & 4 & 4 & 8 & 16 \\ (31) & 0 & 4 & 8 & 16 & 32 \end{array}$$

$$K_5^{(2)} = \begin{array}{c|ccccccc} & (5) & (41) & (32) & (311) & (221) & (2111) & (11111) \\ \hline (5) & 2 & 4 & 4 & 8 & 8 & 16 & 32 \\ (41) & 2 & 8 & 12 & 24 & 32 & 64 & 128 \\ (32) & -2 & 0 & 8 & 16 & 32 & 64 & 128 \end{array}$$

$$K_3^{(3)} = \begin{array}{c|ccc} & (3) & (21) & (111) \\ \hline (3) & 9 & 18 & 27 \\ (21) & 0 & 9 & 27 \end{array}$$

$$K_4^{(3)} = \begin{array}{c|ccccc} & (4) & (31) & (22) & (211) & (1111) \\ \hline (4) & 12 & 27 & 36 & 54 & 81 \\ (31) & 15 & 54 & 72 & 135 & 243 \\ (22) & -3 & 0 & 9 & 27 & 81 \\ (211) & 3 & 27 & 45 & 108 & 243 \end{array}$$

From these numerical experiments, one readily observes that $\kappa_{\lambda\mu}$ is an integral multiple of $p^{\ell(\mu)}$. We do not have a proof for this "fact" at this moment. Here we verify it only for the special case $\lambda = (n)$.

Theorem 4.1. *Let a_n $(n = 0, 1, 2, \ldots)$ be defined by the q-series*

$$\frac{1 - q^p}{(1 - q)^p} = \sum_{n=0}^{\infty} a_n q^n.$$

Then, for $\mu = (\mu_1, \ldots \mu_\ell)$, we have

$$\kappa_{(n)\mu} = a_{\mu_1} a_{\mu_2} \cdots a_{\mu_\ell}.$$

In particular, $\kappa_{(n)\mu}$ is a multiple of $p^{\ell(\mu)}$.

Proof. Putting $y = (q, 0, 0, \ldots)$ in the Cauchy formula (Theorem 3.1), we have

$$\sum_{\lambda \in P^r} \varphi(B^\lambda(x)) \widetilde{B}^\lambda(q, 0, 0, \ldots) = \prod_i \frac{1 - x_i^p q^p}{(1 - x_i q)^p}.$$

Since $\widetilde{B^\lambda}$ is a linear combination of Schur functions, $\widetilde{B^\lambda}(q, 0, 0, \ldots)$ is non-zero only for partitions λ with length at most 1. Therefore the summation in the left-hand side is only over the partitions $\lambda = (n)$ $(n = 0, 1, 2, \ldots)$. And we know that $\widetilde{B^{(n)}}(q, 0, 0, \ldots) = q^n$. Hence we have

$$\sum_{n=0}^{\infty} \varphi(B^{(n)}(x))q^n = \prod_i \frac{1 - x_i^p q^p}{(1 - x_i q)^p}.$$

Here we have

$$\frac{1 - x_i^p q^p}{(1 - x_i q)^p} = \sum_{m_i=0}^{\infty} a_{m_i} x_i^{m_i} q^{m_i}.$$

Taking a product we obtain

$$\prod_i \frac{1 - x_i^p q^p}{(1 - x_i q)^p} = \prod_i \left(\sum_{m_i=0}^{\infty} a_{m_i} x_i^{m_i} q^{m_i} \right)$$

$$= \sum_{n=0}^{\infty} \left(\sum_{\mu \in P(n)} a_{m_1} \cdots a_{m_\ell} m_\mu \right) q^n.$$

as desired.

By definition of a_n, we see that

$$1 - q^p = (1 - q)^p \left(1 + \sum_{n=1}^{\infty} a_n q^n \right)$$

$$\equiv (1 - q^p) \left(1 + \sum_{n=1}^{\infty} a_n q^n \right) \quad (\text{mod } p).$$

It is equivalent to say that

$$\sum_{n=1}^{\infty} a_n q^n \equiv 0 \quad (\text{mod } p),$$

which means that $a_n \equiv 0$ (mod p). Therefore $\kappa_{(n)\mu}$ is a multiple of $p^{\ell(\mu)}$. $\square$

By a simple computation, one sees that

$$\frac{1-q^2}{(1-q)^2} = 1 + \sum_{n=1}^{\infty} 2q^n,$$

$$\frac{1-q^3}{(1-q)^3} = 1 + \sum_{n=1}^{\infty} (3n)q^n,$$

$$\frac{1-q^5}{(1-q)^5} = 1 + 5q + 15q^2 + 35q^3 + 70q^4 + 125q^5$$
$$+ 205q^6 + 315q^7 + 460q^8 + 645q^9 + 875q^{10} + \cdots.$$

References

1. S. Ariki, T. Nakajima and H.-F. Yamada, Reduced Schur functions and Littlewood-Richardson coefficients, J. London Math. Soc. (2) 59 (1999), 396–406.
2. G. D. James and A. Kerber, *The Representation Theory of the Symmetric Groups*, Addison-Wesley, 1979.
3. I. G. Macdonald, *Symmetric Functions and Hall Polynomials, 2nd. edn.*, Oxford, 1995.
4. H. Nagao and Y. Tsushima, *Representations of Finite Groups*, Academic Press, 1989.
5. K. Uno and H.-F. Yamada, Littlewood-Richardson coefficients for Brauer-Schur functions, in preparation.
6. H.-F. Yamada, The basic representation of the extended affine Lie algebra of type $A_1^{(1)}$ and the BKP hierarchy, Lett. Math. Phys. 9 (1985), 133–137.

$\mathcal{O}$-OPERATORS ON ASSOCIATIVE ALGEBRAS, ASSOCIATIVE YANG-BAXTER EQUATIONS AND DENDRIFORM ALGEBRAS

CHENGMING BAI

Chern Institute of Mathematics & LPMC, Nankai University, Tianjin 300071, China
E-mail: baicm@nankai.edu.cn

LI GUO

Department of Mathematics and Computer Science,
Rutgers University, Newark, NJ 07102
E-mail: liguo@rutgers.edu

XIANG NI

Chern Institute of Mathematics & LPMC, Nankai University, Tianjin 300071, China
E-mail: xiangn_math@yahoo.cn

This paper studies $\mathcal{O}$-operators and the role they play in relating diverse objects such as Rota-Baxter operators, associative Yang-Baxter equations and dendriform algebras. The paper gives a survey of recent developments. It also presents new results on antisymmetric infinitesimal bialgebras, discusses the relationship of $\mathcal{O}$-operators with Rota-Baxter operators and relative differential operators, provides further characterizations of dendriform algebras and constructs solutions of associative Yang-Baxter equations from dendriform algebras.

Keywords: $\mathcal{O}$-operators, associative Yang-Baxter equations, dendriform algebras, infinitesimal bialgebras, differential operators.

1. Introduction

In this paper we give a survey of some recent developments[8-10] on $\mathcal{O}$-operators and their relationship with Rota-Baxter operators, associative Yang-Baxter equations and dendriform algebras. We also present new results on antisymmetric infinitesimal bialgebras and give further characterizations of dendriform algebras. Furthermore, the relationship of $\mathcal{O}$-operators with Rota-Baxter operators and relative differential operators is considered.

Rota-Baxter operators, Yang-Baxter equations and dendriform algebras are quite distinct objects. Not only they have distinct origins, with Rota-Baxter algebras arising from probability and combinatorics, Yang-Baxter equations arising from integrable systems and dendriform algebras arising from K-theory and operads, they also have distinct forms, with Rota-Baxter operator a type of linear operator, Yang-Baxter equation a certain tensor equation and dendriform algebra an algebraic structure with multiple binary multiplications. Their connections were first found by Aguiar[1,2] who showed that a solution of the associative Yang-Baxter equation gives rise to a Rota-Baxter operator of weight zero and a Rota-Baxter operator of weight zero in turn gives rise to a dendriform dialgebra. Such connections have been pursued further in several subsequent papers.[3,5,6,16,18] But it is desirable to find precise connections among the three objects.

In this paper, we show that these connections can be understood better upon generalizing the concept of a Rota-Baxter operator to that of an $\mathcal{O}$-operator. It is a relative version of the Rota-Baxter operator and, in the context of Lie algebras, was first defined by Kupershmidt[25] in the study of Yang-Baxter equations even though the same structure was used by Bordemann[13] in the study of integrable systems. We generalize the aforementioned relations from Rota-Baxter operators to $\mathcal{O}$-operators. This generalization also allows us to understand better these relations.

1.1. *Rota-Baxter algebras, Yang-Baxter equations and dendriform algebras*

We recall concepts and relations that motivated our study.

To fix notations, we let **k** denote a field in this paper even though most of the results hold when **k** is a commutative unitary ring. By a **k**-algebra we mean an associative (not necessarily unitary) **k**-algebra, unless otherwise stated.

Definition 1.1. Let R be a **k**-algebra and let $\lambda \in \mathbf{k}$ be given. If a **k**-linear map $P : R \to R$ satisfies the **Rota-Baxter relation**:

$$P(x)P(y) = P(P(x)y) + P(xP(y)) + \lambda P(xy), \quad \forall x, y \in R, \qquad (1)$$

then P is called a **Rota-Baxter operator of weight** λ and (R, P) is called a **Rota-Baxter algebra of weight** λ.

The study of Rota-Baxter algebras originated from probability and combinatorics in the 1950s and 1960s,[11,14,30,31] and have found broad applications since late 1990s in mathematical physics, number theory and operads.[2,3,12,15,18–20,22–24,26]

For simplicity, we will only discuss the case of Rota-Baxter operators of weight zero in the introduction.

Note that the relation (1) still makes sense when R is replaced by a k-vector space with a binary operation such as the Lie bracket, giving rise to the concept of a Rota-Baxter operator on a Lie algebra. If the Lie algebra is equipped with a nondegenerate symmetric invariant bilinear form, then a skew-symmetric solution of the **classical Yang-Baxter equation (CYBE)**

$$[r_{12}, r_{13}] + [r_{12}, r_{23}] + [r_{13}, r_{23}] = 0$$

is just a Rota-Baxter operator of weight zero. We refer the reader to[5,32] and later sections for further details.

Motivated by the importance of CYBE, we consider the following associative analogue.

Definition 1.2. ([1–3]) Let A be a k-algebra. An element $r \in A \otimes A$ is called a **solution of the associative Yang-Baxter equation in A** if it satisfies the relation

$$r_{12}r_{13} + r_{13}r_{23} - r_{23}r_{12} = 0, \tag{2}$$

called the **associative Yang-Baxter equation (AYBE)**. Here, for $r = \sum_i a_i \otimes b_i \in A \otimes A$, denote

$$r_{12} = \sum_i a_i \otimes b_i \otimes 1, \quad r_{13} = \sum_i a_i \otimes 1 \otimes b_i, \quad r_{23} = \sum_i 1 \otimes a_i \otimes b_i. \tag{3}$$

Also we equip $A \otimes A \otimes A$ with the product of the tensor algebra. In particular,

$$(a_1 \otimes a_2 \otimes a_3)(b_1 \otimes b_2 \otimes b_3) = (a_1 b_1) \otimes (a_2 b_2) \otimes (a_3 b_3), \quad \forall a_i, b_i \in A, i = 1, 2, 3.$$

In the opposite algebra of A, Eq. (2) takes the form[1–3]

$$r_{13}r_{12} - r_{12}r_{23} + r_{23}r_{13} = 0. \tag{4}$$

As an associative analogue of the relationship between Rota-Baxter operators on Lie algebras and CYBE, Aguiar[1,2] showed that, for a solution $r = \sum_i a_i \otimes b_i \in A \otimes A$ of Eq. (4) in A, the map

$$P : A \to A, \quad P(x) = \sum_i a_i x b_i, \quad \forall x \in A,$$

defines a Rota-Baxter operator of weight zero on A.

On the other hand, with motivation from periodicity of algebraic K-theory and operads, dendriform dialgebras were introduced by Loday[27] in the 1990s.

Definition 1.3. A **dendriform dialgebra** is a triple $(R, \prec, \succ)$ consisting of a k-vector space R and two binary operations $\prec$ and $\succ$ on R such that

$$(x \prec y) \prec z = x \prec (y \star z), \quad (x \succ y) \prec z = x \succ (y \prec z),$$
$$x \succ (y \succ z) = (x \star y) \succ z, \tag{5}$$

for all $x, y, z \in R$. Here $x \star y = x \prec y + x \succ y$.

Aguiar[1] first established that, for a Rota-Baxter k-algebra (R, P) of weight zero, the binary operations

$$x \prec_P y = xP(y), \quad x \succ_P y = P(x)y, \quad \forall x, y \in R, \tag{6}$$

define a dendriform dialgebra $(R, \prec_P, \succ_P)$.

This defines a functor from the category of Rota-Baxter algebras of weight 0 to the category of dendriform dialgebras. This work has inspired quite a few subsequent studies[4,6,7,12,16,18,21] that generalized and further clarified the relationship between Rota-Baxter algebras and dendriform dialgebras and trialgebras of Loday and Ronco,[28] including the adjoint functor of the above functor, the related Poincare-Birkhoff-Witt theorem and Gröbner-Shirshov basis.

To summarize, for a given k-vector space R, let $\mathrm{RB}(R)$ denote the set of Rota-Baxter operators on R, $\mathrm{AYB}(R)$ denote the set of solutions of AYBE in R and $\mathrm{DD}(R)$ denote the set of dendriform dialgebra structures on R (namely the set of pairs $(\prec, \succ)$ of binary operations on R that satisfy Eq. (5)). Then we obtain the maps

$$\mathrm{AYB}(R) \longrightarrow \mathrm{RB}(R) \longrightarrow \mathrm{DD}(R) \tag{7}$$

These studies suggested that there should be a close relationship between Rota-Baxter algebras, AYBE and dendriform dialgebras. In particular, one can try to reverse the above maps and ask whether a Rota-Baxter operator can give a solution of AYBE, or whether every dendriform dialgebra and trialgebra could be derived from a Rota-Baxter algebra by a construction like Eq. (6). As later examples show, this is quite far from being the case. Thus a suitable generalization of Rota-Baxter operators is needed. The purpose of this paper is to show that such a generalization is given by the concept of (extended) $\mathcal{O}$-operators motivated by such a concept on Lie algebras.

1.2. $\mathcal{O}$-operators and layout of the paper

In this paper, we introduce the concept of an extended $\mathcal{O}$-operator as a generalization of the concept of a Rota-Baxter operator and the associative analogue of an $\mathcal{O}$-operator on a Lie algebra.

Then on one hand, we extend the connections of Rota-Baxter algebras with AYBE to those of $\mathcal{O}$-operators. This study is motivated by the relationship between $\mathcal{O}$-operator and CYBE in Lie algebras.[5,8,13,25] On the other hand, to clarify the connection between Rota-Baxter operators and dendriform dialgebras and trialgebras, we show that an $\mathcal{O}$-operator on a k-vector space (resp. a k-algebra) could derive all the dendriform dialgebras (resp. trialgebras).

Moreover, we prove new results on the relationship of AYBE with antisymmetric infinitesimal bialgebras and dendriform algebras by making use of the connection obtained above of $\mathcal{O}$-operators with AYBE and dendriform algebras. We also obtain characterizations of dendriform algebras and consider the relationship of $\mathcal{O}$-operators with Rota-Baxter operators and relative differential operators.

Here is the layout of the paper. In Section 2, the concept of an extended $\mathcal{O}$-operator is introduced and its first connection with Rota-Baxter operators is studied. Section 3 establishes the two way relationship of extended $\mathcal{O}$-operators with the AYBE and the extended AYBE. In Section 4, we first discuss the relationship between generalized AYBE and extended $\mathcal{O}$-operators and then consider the double space of a factorizable quasitriangular infinitesimal bialgebra. In Section 5, we derive all dendriform algebra structures from $\mathcal{O}$-operators. In the final Section 6, we first return to the relationship between $\mathcal{O}$-operators and Rota-Baxter operators. We also introduced the concept of a relative differential operator as a differential variation of an $\mathcal{O}$-operator, and study its relationship with various forms of AYBE. We finally combine results from Section 3 and Section 5 to establish direct connections between various forms of AYBE and dendriform algebras.

Section 2, Section 3 and Section 4.1 are based on[9] while Section 5 is based on.[10] Results in Section 4.2 and Section 6 are new.

2. $\mathcal{O}$-operators and extended $\mathcal{O}$-operators

We give background notations in Section 2.1 before introducing the concept of an extended $\mathcal{O}$-operator in Section 2.2. We then show in Section 2.3 that certain extended $\mathcal{O}$-operators on an algebra can be derived from Rota-

Baxter operators on another algebra with the same underlying vector space. The main reference of this section is.[9]

2.1. *Bimodules and A-bimodule k-algebras*

We first recall the concept of a bimodule.

Definition 2.1. Let $(A, \cdot)$ be a k-algebra.

(a) An **A-bimodule** is a k-vector space V, together with linear maps $\ell, r : A \to \mathrm{End}_{\mathbf{k}}(V)$, such that (V, ℓ) defines a left A-module, (V, r) defines a right A-module and the two module structures on V are compatible in the sense that

$$(\ell(x)v)r(y) = \ell(x)(vr(y)), \ \forall \, x, y \in A, v \in V.$$

If we want to be more precise, we also denote an A-bimodule V by the triple (V, ℓ, r).

(b) A **homomorphism** between two A-bimodules (V_1, ℓ_1, r_1) and (V_2, ℓ_2, r_2) is a k-linear map $g : V_1 \to V_2$ such that

$$g(\ell_1(x)v) = \ell_2(x)g(v), \quad g(vr_1(x)) = g(v)r_2(x), \quad \forall \, x \in A, v \in V_1.$$

For a k-algebra A and $x \in A$, define the left and right actions

$$L(x) : A \to A, \ L(x)y = xy; \quad R(x) : A \to A, \ yR(x) = yx, \quad y \in A.$$

Further define

$$L = L_A : A \to \mathrm{End}_{\mathbf{k}}(A), \ x \mapsto L(x); \quad R = R_A : A \to \mathrm{End}_{\mathbf{k}}(A), \ x \mapsto R(x),$$

$x \in A$.

Obviously, (A, L, R) is an A-bimodule.

For a k-vector space V, let $V^* := \mathrm{Hom}_{\mathbf{k}}(V, \mathbf{k})$ denote the dual k-vector space. Denote the usual pairing between V^* and V by

$$\langle \, , \, \rangle : V^* \times V \to \mathbf{k}, \langle u^*, v \rangle = u^*(v), \forall u^* \in V^*, v \in V.$$

Let A be a k-algebra and let (V, ℓ, r) be an A-bimodule. Define the linear maps $\ell^*, r^* : A \to \mathrm{End}_{\mathbf{k}}(V^*)$ by

$$\langle u^*\ell^*(x), v \rangle = \langle u^*, \ell(x)v \rangle, \langle r^*(x)u^*, v \rangle = \langle u^*, vr(x) \rangle, \forall x \in A, u^* \in V^*, v \in V,$$

respectively. Then (V^*, r^*, ℓ^*) is an A-bimodule, called the **dual bimodule**[6] of (V, ℓ, r).

Let (A^*, R^*, L^*) denote the dual A-bimodule of the A-bimodule (A, L, R).

We next extend the concept of an A-bimodule to that of an A-bimodule k-algebra by replacing the k-vector space V by a k-algebra R.

Definition 2.2. Let $(A, \cdot)$ be a k-algebra with multiplication $\cdot$.

(a) Let $(R, \circ)$ be a k-algebra with multiplication $\circ$. Let $\ell, r : A \to \mathrm{End}_\mathbf{k}(R)$ be two linear maps. We call R (or the triple (R, ℓ, r) or the quadruple $(R, \circ, \ell, r)$) an A-**bimodule k-algebra** if (R, ℓ, r) is an A-bimodule that is compatible with the multiplication $\circ$ on R. More precisely, we have

$$\ell(x \cdot y)v = \ell(x)(\ell(y)v), \ \ell(x)(v \circ w) = (\ell(x)v) \circ w,$$
$$vr(x \cdot y) = (vr(x))r(y), \ (v \circ w)r(x) = v \circ (wr(x)),$$
$$(\ell(x)v)r(y) = \ell(x)(vr(y)), \ (vr(x)) \circ w = v \circ (\ell(x)w), \ \forall \, x, y \in A, v, w \in R.$$

(b) A **homomorphism** between two A-bimodule k-algebras $(R_1, \circ_1, \ell_1, r_1)$ and $(R_2, \circ_2, \ell_2, r_2)$ is a k-linear map $g : R_1 \to R_2$ that is both an A-bimodule homomorphism and a k-algebra homomorphism.

The reader should be alerted that an A-bimodule k-algebra R need not be a left or right A-algebra since we do not assume that $A \cdot 1$ is in the center of R. For example, the A-bimodule (A, L, R) is an A-bimodule k-algebra with the product of A. But it is not an A-algebra unless A is a commutative ring.

For the more general concept of a matched pair, see[6] and[9] from which we also cite the following result.

Proposition 2.3. *Let $(A, \cdot)$ and $(R, \circ)$ be two k-algebras. Let $\ell, r : A \to \mathrm{End}_\mathbf{k}(R)$ be two linear maps. Then $(R, \circ, \ell, r)$ is an A-bimodule k-algebra if and only if the direct sum $A \oplus R$ of vector spaces is turned into a k-algebra (the semidirect sum) by defining a multiplication on $A \oplus R$ by*

$$(x_1, v_1) * (x_2, v_2) = (x_1 \cdot x_2, \ell(x_1)v_2 + v_1 r(x_2) + v_1 \circ v_2), \ \forall x_1, x_2 \in A, v_1, v_2 \in R. \tag{8}$$

We denote this algebra by $A \ltimes_{\ell,r} R$ or simply $A \ltimes R$.

2.2. *Extended $\mathcal{O}$-operators*

We first define an $\mathcal{O}$-operator before introducing an extended $\mathcal{O}$-operator through an auxiliary operator.

2.2.1. $\mathcal{O}$-operators

Definition 2.4. Let $(A, \cdot)$ be a k-algebra and $(R, \circ, \ell, r)$ be an A-bimodule k-algebra. A linear map $\alpha : R \to A$ is called an $\mathcal{O}$-**operator of weight** $\lambda \in \mathbf{k}$ **associated to** $(R, \circ, \ell, r)$ if α satisfies

$$\alpha(u) \cdot \alpha(v) = \alpha(\ell(\alpha(u))v) + \alpha(ur(\alpha(v))) + \lambda\alpha(u \circ v), \quad \forall u, v \in V. \quad (9)$$

Obviously, an $\mathcal{O}$-operator associated to (A, L, R) is just a Rota-Baxter operator on A. An $\mathcal{O}$-operator can be viewed as the relative version of a Rota-Baxter operator in the sense that the domain and range of an $\mathcal{O}$-operator might be different.

Note that, an A-bimodule (V, ℓ, r) becomes an A-bimodule k-algebra when V is equipped with the zero multiplication. Then a linear map $\alpha : V \to A$ is an $\mathcal{O}$-operator (of any weight λ) if

$$\alpha(u) \cdot \alpha(v) = \alpha(\ell(\alpha(u))v) + \alpha(ur(\alpha(v))), \quad \forall u, v \in V.$$

Since the weight λ makes no difference in the definition, we just call α an $\mathcal{O}$-operator. Such a structure appeared independently in[33] under the name of a **generalized Rota-Baxter operator**.

2.2.2. Balanced homomorphisms

For our purpose of further generalizing the concept of an $\mathcal{O}$-operator, we introduce another concept.

Definition 2.5. Let $(A, \cdot)$ be a k-algebra.

(a) Let $\kappa \in \mathbf{k}$ and let (V, ℓ, r) be an A-bimodule. A linear map (resp. an A-bimodule homomorphism) $\beta : V \to A$ is called a **balanced linear map of mass** κ (resp. **balanced A-bimodule homomorphism of mass** κ) if

$$\kappa\ell(\beta(u))v = \kappa ur(\beta(v)), \quad \forall u, v \in V. \quad (10)$$

(b) Let $\kappa, \mu \in \mathbf{k}$ and let $(R, \circ, \ell, r)$ be an A-bimodule k-algebra. A linear map (resp. an A-bimodule homomorphism) $\beta : R \to A$ is called a **balanced linear map of mass** (κ, μ) (resp. a **balanced A-bimodule homomorphism of mass** (κ, μ)) if Eq. (10) holds and

$$\mu\ell(\beta(u \circ v))w = \mu ur(\beta(v \circ w)), \quad \forall u, v, w \in R. \quad (11)$$

Under our assumption that **k** is a field, we only need to consider the cases when the values of κ and μ are taken from $\{0,1\}$. We take κ and μ to be parameters from **k** in order to give a uniform treatment of these different cases and to work with a general base ring **k**.

2.2.3. *Extended $\mathcal{O}$-operators*

We can now introduce our main concept in this paper.

Definition 2.6. Let $(A, \cdot)$ be a **k**-algebra and let $(R, \circ, \ell, r)$ be an A-bimodule **k**-algebra.

(a) Let $\lambda, \kappa, \mu \in$ **k**. Fix a balanced A-bimodule homomorphism β: $(R, \ell, r) \to A$ of mass (κ, μ). A linear map $\alpha : R \to A$ is called an **extended $\mathcal{O}$-operator of weight λ with modification β of mass** (κ, μ) if

$$\alpha(u) \cdot \alpha(v) - \alpha(\ell(\alpha(u))v + ur(\alpha(v)) + \lambda u \circ v) = \kappa\beta(u) \cdot \beta(v) + \mu\beta(u \circ v), \tag{12}$$

for all $u, v \in R$.

(b) We also let (α, β) denote an extended $\mathcal{O}$-operator α with modification β.

(c) When (V, ℓ, r) is an A-bimodule, we regard V as an A-bimodule **k**-algebra with the zero multiplication. Then λ and μ are irrelevant. We then call the pair (α, β) an **extended $\mathcal{O}$-operator with modification β of mass κ.**

We note that, when the modification β is the zero map (and hence κ and μ are irrelevant), then α is the $\mathcal{O}$-operator defined in Definition 2.4.

2.3. *Extended $\mathcal{O}$-operators, $\mathcal{O}$-operators and Rota-Baxter operators: the first connection*

This section studies the relationship between extended $\mathcal{O}$-operators, $\mathcal{O}$-operators and Rota-Baxter operators. Further results on this relationship will be given in Section 6.1.

Let $(A, \cdot)$ be a **k**-algebra and $(R, \circ, \ell, r)$ be an A-bimodule **k**-algebra. The next two results show how an extended $\mathcal{O}$-operator can be regarded as a (non-extended) $\mathcal{O}$-operator by a change of the product on the underlying vector space.

Theorem 2.7. *Let $(A, \cdot)$ be a **k**-algebra and $(R, \circ, \ell, r)$ be an A-bimodule* **k**-*algebra. Let $\alpha, \beta : R \to A$ be two linear maps and $\lambda \in \mathbf{k}$. Suppose β is a balanced A-bimodule homomorphism of mass $(-1, \pm\lambda)$, that is, β satisfies Eq. (10) with $\kappa = -1$, Eq. (11) with $\mu = \pm\lambda$ and*

$$\beta(\ell(x)u) = x \cdot \beta(u), \quad \beta(ur(x)) = \beta(u) \cdot x, \quad \forall x \in A, u \in R.$$

Define products $\circ_{\pm} = \circ_{\lambda, \beta, \pm}$ on R by

$$u \circ_{\pm} v = \lambda u \circ v \mp 2\ell(\beta(u))v, \quad \forall u, v \in R.$$

*Then α is an extended $\mathcal{O}$-operator of weight λ with modification β of mass $(\kappa, \mu) = (-1, \pm\lambda)$ if and only if $\alpha \pm \beta$ are $\mathcal{O}$-operators of weight 1 associated to the A-bimodule **k**-algebra $(R, \circ_{\pm}, \ell, r)$:*

$$(\alpha\pm\beta)(u)\cdot(\alpha\pm\beta)(v) = (\alpha\pm\beta)(\ell((\alpha\pm\beta)(u))v+ur((\alpha\pm\beta)(v))+u\circ_{\pm}v), \quad (13)$$

for all $u, v \in R$.

By taking R to be a vector space over **k** with the zero product, we obtain:

Corollary 2.8. *Let A be a **k**-algebra and (V, ℓ, r) be an A-bimodule. Let $\alpha, \beta : V \to A$ be two linear maps such that β is a balanced A-bimodule homomorphism. Define products $\star_{\pm}$ on V by*

$$u \star_{\pm} v = \mp 2\ell(\beta(u))v, \quad \forall u, v \in V.$$

*Then α is an extended $\mathcal{O}$-operator with modification β of mass $\kappa = -1$ if and only if $\alpha \pm \beta$ is an $\mathcal{O}$-operator of weight 1 associated to the A-bimodule **k**-algebra $(V, \star_{\pm}, \ell, r)$.*

We next consider the special case when the A-bimodule **k**-algebra is taken to be $(A, \cdot, L, R)$. In the case of $\mu = 0$, Eq. (12) takes the form

$$\alpha(x) \cdot \alpha(y) - \alpha(\alpha(x) \cdot y + x \cdot \alpha(y) + \lambda x \cdot y) = \kappa\beta(x) \cdot \beta(y), \quad \forall x, y \in A. \quad (14)$$

We list the following special cases for later reference. When $\lambda = 0$, Eq. (14) gives

$$\alpha(x) \cdot \alpha(y) - \alpha(\alpha(x) \cdot y + x \cdot \alpha(y)) = \kappa\beta(x) \cdot \beta(y), \quad \forall x, y \in A. \quad (15)$$

If in addition, $\beta = \mathrm{id}$, then Eq. (15) gives

$$\alpha(x) \cdot \alpha(y) - \alpha(\alpha(x) \cdot y + x \cdot \alpha(y)) = \kappa x \cdot y, \quad \forall x, y \in A. \quad (16)$$

When furthermore $\kappa = -1$, Eq. (16) becomes

$$\alpha(x) \cdot \alpha(y) - \alpha(\alpha(x) \cdot y + x \cdot \alpha(y)) = -x \cdot y, \quad \forall x, y \in A. \quad (17)$$

The following results reduce extended $\mathcal{O}$-operators to $\mathcal{O}$-operators and Rota-Baxter operators by changing the product on the underlying vector space.

Corollary 2.9. *Let $(A, \cdot)$ be a $\mathbf{k}$-algebra and let $\alpha, \beta : A \to A$ be linear maps. Suppose β is an A-bimodule homomorphism and define products $\star_\pm$ on A by*

$$x \star_\pm y = \mp 2\beta(x) \cdot y, \quad \forall x, y \in A.$$

Then α and β satisfy Eq. (15) for $\kappa = -1$ if and only if $\alpha \pm \beta$ are $\mathcal{O}$-operators of weight 1 associated to the A-bimodule $\mathbf{k}$-algebra $(A, \star_\pm, L, R)$.

Let $(A, \cdot)$ be a $\mathbf{k}$-algebra and let $(A, \cdot, L, R)$ be the corresponding A-bimodule $\mathbf{k}$-algebra. In this case, it is obvious that $\beta = \mathrm{id}$ satisfies the conditions of Theorem 2.7 and Eq. (12) takes the form

$$\alpha(x) \cdot \alpha(y) - \alpha(\alpha(x) \cdot y + x \cdot \alpha(y) + \lambda x \cdot y) = \hat{\kappa} x \cdot y, \quad \forall x, y \in A, \quad (18)$$

where $\hat{\kappa} = \kappa + \mu$. Thus we have the following consequence of Theorem 2.7.

Corollary 2.10. *Let $\hat{\kappa} = -1 \pm \lambda$. Then $\alpha : A \to A$ satisfies Eq. (18) if and only if $\alpha \pm 1$ is a Rota-Baxter operator of weight $\lambda \mp 2$.*

When λ is invertible, we get the following conclusion.

Corollary 2.11. ([16,32]) *α is a Rota-Baxter operator of weight $\lambda \neq 0$ if and only if $\frac{2}{\lambda}\alpha + \mathrm{id}$ is an extended $\mathcal{O}$-operator with modification id of mass -1.*

3. Extended $\mathcal{O}$-operators and AYBE

In this section we study the relationship between extended $\mathcal{O}$-operators and AYBE. We start with introducing a generalization of AYBE in Section 3.1. We then establish connections between $\mathcal{O}$-operators in different generalities and solutions of these variations of AYBE in different algebras. The relationship between $\mathcal{O}$-operators on a $\mathbf{k}$-algebra A and solutions of AYBE in A is considered in Section 3.2, together with the special case of Frobenius algebras. We consider in Section 3.3 the relationship between extended $\mathcal{O}$-operators and solutions of AYBE and extended AYBE in an extension algebra of A.

3.1. *Extended AYBE*

We define variations of AYBE to be satisfied by 2-tensors from an algebra. We then study linear maps from these two tensors in preparation to study the relationship between $\mathcal{O}$-operators and solutions of these variations of AYBE.

Let A be a **k**-algebra. Let $r = \sum_i a_i \otimes b_i \in A \otimes A$. We continue to use the notations r_{12}, r_{13} and r_{23} defined in Eq. (3). We similarly define

$$r_{21} = \sum_i b_i \otimes a_i \otimes 1, \quad r_{31} = \sum_i b_i \otimes 1 \otimes a_i, \quad r_{32} = \sum_i 1 \otimes b_i \otimes a_i.$$

Definition 3.1. Fix an $\varepsilon \in \mathbf{k}$.

(a) The equation

$$r_{12}r_{13} + r_{13}r_{23} - r_{23}r_{12} = \varepsilon(r_{13} + r_{31})(r_{23} + r_{32}) \tag{19}$$

is called the **extended associative Yang-Baxter equation of mass ε** (or ε-**EAYBE** in short).

(b) Let A be a **k**-algebra. An element $r \in A \otimes A$ is called a **solution of the ε-EAYBE in A** if it satisfies Eq. (19).

When $\varepsilon = 0$ or when r is **skew-symmetric** in the sense that $\sigma(r) = -r$ for the switch operator

$$\sigma : A \otimes A \to A \otimes A, \quad \sigma\left(\sum_i a_i \otimes b_i\right) = \sum_i b_i \otimes a_i$$

(and hence $r_{13} = -r_{31}$), then the ε-EAYBE is the same as the AYBE in Eq. (2):

$$r_{12}r_{13} + r_{13}r_{23} - r_{23}r_{12} = 0. \tag{20}$$

Let

$$t : \mathrm{Hom}_{\mathbf{k}}(A^*, A) \to \mathrm{Hom}_{\mathbf{k}}(A^*, A)$$

be the transpose operator that sends $\alpha : A^* \to A$ to $\alpha^t : A^* \to A$, characterized by

$$\langle u, \alpha^t(v) \rangle = \langle \alpha(u), v \rangle, \quad u, v \in A^*.$$

When A is finite-dimensional over **k**, the natural bijection

$$\phi : A \otimes A \to \mathrm{Hom}_{\mathbf{k}}(A^*, \mathbf{k}) \otimes A \to \mathrm{Hom}_{\mathbf{k}}(A^*, A),$$
$$\langle v, \phi(r)(u) \rangle = \langle u \otimes v, r \rangle, \forall u, v \in A^*,$$

is compatible with the operators σ and t. Let $\mathrm{Sym}^2(A \otimes A)$ and $\mathrm{Alt}^2(A \otimes A)$ (resp. $\mathrm{Hom}_{\mathbf{k}}(A^*, A)_+$ and $\mathrm{Hom}_{\mathbf{k}}(A^*, A)_-$) be the eigenspaces for the eigenvalues 1 and -1 of σ on $A \otimes A$ (resp. of t on $\mathrm{Hom}_{\mathbf{k}}(A^*, A)$). Thus we have the commutative diagram of bijective linear maps:

$$
\begin{array}{ccc}
A \otimes A & \xrightarrow{\quad \phi \quad} & \mathrm{Hom}_{\mathbf{k}}(A^*, A) \\
\downarrow & & \downarrow \\
\mathrm{Alt}^2(A \otimes A) \oplus \mathrm{Sym}^2(A \otimes A) & \xrightarrow{\;\phi\;} & \mathrm{Hom}_{\mathbf{k}}(A^*, A)_- \oplus \mathrm{Hom}_{\mathbf{k}}(A^*, A)_+
\end{array}
\tag{21}
$$

preserving the factorizations. We will often denote $\phi(r)$ by r for simplicity of notations.

Suppose that the characteristic of $\mathbf{k}$ is not 2 and define

$$
\alpha = \alpha_r = (r - r^t)/2, \quad \beta = \beta_r = (r + r^t)/2,
\tag{22}
$$

called the **skew-symmetric part** and the **symmetric part** of r respectively. Then $r = \alpha + \beta$ and $r^t = -\alpha + \beta$.

Definition 3.2. Let $(A, \cdot)$ be a **k**-algebra. Let $s \in A \otimes A$ be symmetric. If

$$
(\mathrm{id} \otimes L(x) - R(x) \otimes \mathrm{id})s = 0, \quad \forall x \in A,
\tag{23}
$$

then s is called **invariant**.

We give the following equivalent conditions for the invariant property.

Proposition 3.3. *Let $(A, \cdot)$ be a* **k**-*algebra with finite* **k**-*dimension. Let $s \in A \otimes A$ be symmetric. Then s is invariant if and only if one of the following two conditions holds:*

(a) s regarded as a linear map from (A^*, R^*, L^*) to A is balanced, i.e.,

$$
R^*(s(a^*))b^* = a^* L^*(s(b^*)), \quad \forall a^*, b^* \in A^*.
$$

(b) s regarded as a linear map from (A^*, R^*, L^*) to A is an A-bimodule homomorphism, i.e.,

$$
s(R^*(x)a^*) = x \cdot s(a^*), \quad s(a^* L^*(x)) = s(a^*) \cdot x, \quad \forall x \in A, a^* \in A^*.
$$

3.2. *From EAYBE to Extended $\mathcal{O}$-operators*

The following theorem shows how a solution of AYBE in an algebra A can be characterized in terms of an extended $\mathcal{O}$-operators on A.

Theorem 3.4. *Let $\mathbf{k}$ be a field of characteristic not equal to 2. Let A be a $\mathbf{k}$-algebra with finite $\mathbf{k}$-dimension and let $r \in A \otimes A$ which is identified as a linear map from A^* to A.*

(a) *Then r is a solution of AYBE in A if and only if r satisfies*

$$r(a^*) \cdot r(b^*) = r(R^*(r(a^*))b^* - a^*L^*(r^t(b^*))), \quad \forall a^*, b^* \in A^*.$$

(b) *Define α and β by Eq. (22). Suppose that the symmetric part β of r is invariant. Then r is a solution of EAYBE of mass $\frac{\kappa+1}{4}$:*

$$r_{12}r_{13} + r_{13}r_{23} - r_{23}r_{12} = \frac{\kappa+1}{4}(r_{13} + r_{31})(r_{23} + r_{32})$$

if and only if α is an extended $\mathcal{O}$-operator with modification β of mass κ:

$$\alpha(a^*) \cdot \alpha(b^*) - \alpha(R^*(\alpha(a^*))b^* + a^*L^*(\alpha(b^*))) = \kappa\beta(a^*) \cdot \beta(b^*), \quad \forall a^*, b^* \in A^*.$$

In the case of $\kappa = -1$, we have

Corollary 3.5. *Let $\mathbf{k}$ be a field of characteristic not equal to 2. Let A be a $\mathbf{k}$-algebra with finite $\mathbf{k}$-dimension and let $r \in A \otimes A$. Define α and β by Eq. (22).*

(a) *If β is invariant, then the following conditions are equivalent.*

    (i) *r is a solution of AYBE in A.*

    (ii) *The map r (resp. $-r^t$) is an $\mathcal{O}$-operator of weight 1 associated to a new A-bimodule $\mathbf{k}$-algebra $(A^*, \circ_+, R^*, L^*)$ (resp. $(A^*, \circ_-, R^*, L^*)$):*

$$r(a^*) \cdot r(b^*) = r(R^*(r(a^*))b^* + a^*L^*(r(b^*)) + a^* \circ_+ b^*), \quad \forall a^*, b^* \in A^*,$$

(resp.

$$(-r^t)(a^*) \cdot (-r^t)(b^*) = (-r^t)(R^*((-r^t)(a^*))b^* + a^*L^*((-r^t)(b^*))$$
$$+ a^* \circ_- b^*),$$

$\forall a^*, b^* \in A^*$), where the associative products $\circ_\pm$ on A^* are defined by

$$a^* \circ_\pm b^* = \mp 2R^*(\beta(a^*))b^*, \quad \forall a^*, b^* \in A^*.$$

(iii) For any $a^*, b^* \in A^*$,

$$(\alpha \pm \beta)(a^* * b^*) = (\alpha \pm \beta)(a^*) \cdot (\alpha \pm \beta)(b^*), \qquad (24)$$

where

$$a^* * b^* = R^*(r(a^*))b^* - a^*L^*(r^t(b^*)), \quad \forall a^*, b^* \in A^*.$$

(b) When r is skew-symmetric, then r is a solution of AYBE in A if and only if $r : A^* \to A$ is an $\mathcal{O}$-operator of weight zero.

Let $\mathrm{Hom}_{bim}(A^*, A)_+$ be the subset of $\mathrm{Hom}_{\mathbf{k}}(A^*, A)_+$ consisting of A-bimodule homomorphisms from A^* to A both of which are equipped with the natural A-bimodule structures. Denote $\mathrm{Sym}_{bim}^2(A \otimes A) := \phi^{-1}(\mathrm{Hom}_{bim}(A^*, A)_+) \subseteq \mathrm{Sym}^2(A \otimes A)$. Then Corollary 3.5 can be restated as

Proposition 3.6. *An element $r = (r_-, r_+) \in Alt^2(A \otimes A) \oplus Sym_{bim}^2(A \otimes A)$ is a solution of the AYBE in Eq. (2) if and only if the pair $\phi(r) = (\phi(r)_-, \phi(r)_+) = (\phi(r_-), \phi(r_+))$ is an extended $\mathcal{O}$-operator with modification $\phi(r_+)$ of mass $\kappa = -1$. In particular, when r_+ is zero, an element $r = (r_-, 0) = r_- \in Alt^2(A \otimes A)$ is a solution of AYBE if and only if the pair $\phi(r) = (\phi(r)_-, 0) = \phi(r_-)$ is an $\mathcal{O}$-operator of weight zero given by Eq. (9) when (V, ℓ, r) is the dual bimodule (A^*, R^*, L^*) of (A, L, R).*

Let $\mathcal{MO}(A^*, A)$ denote the set of extended $\mathcal{O}$-operators (α, β) from A^* to A of mass $\kappa = -1$. Let $\mathcal{O}(A^*, A)$ denote the set of $\mathcal{O}$-operators $\alpha : A^* \to A$ of weight 0. Let $\mathrm{AYB}(A)$ denote the set of solutions of AYBE (2) in A. Let $\mathrm{SAYB}(A)$ denote the set of skew-symmetric solutions of AYBE (2) in A. Then Proposition 3.6 means that the bijection in Eq. (21) restricts to bijections in the following commutative diagram.

3.3. *From extended $\mathcal{O}$-operators to EAYBE*

In this section we describe an extended $\mathcal{O}$-operator $\alpha : V \to A$ in terms of a solution of EAYBE, especially in the case when A is a Frobenius algebra.

3.3.1. *The general case*

For this purpose we prove that an extended $\mathcal{O}$-operator $\alpha : V \to A$ naturally gives rise to an extended $\mathcal{O}$-operator on a larger k-algebra $\hat{A}$ associated to the dual bimodule $(\hat{A}^*, R_{\hat{A}}^*, L_{\hat{A}}^*)$. We first introduce some notations.

Definition 3.7. Let A be a k-algebra and let (V, ℓ, r) be an A-bimodule, both with finite k-dimensions. Let (V^*, r^*, ℓ^*) be the dual A-bimodule and let $\hat{A} = A \ltimes_{r^*, \ell^*} V^*$. Identify a linear map $\gamma : V \to A$ as an element in $\hat{A} \otimes \hat{A}$ through the injective map

$$\mathrm{Hom_k}(V, A) \cong A \otimes V^* \hookrightarrow \hat{A} \otimes \hat{A}.$$

Denote

$$\tilde{\gamma}_{\pm} := \gamma \pm \gamma^{21}, \tag{25}$$

where $\gamma^{21} = \sigma(\gamma) \in V^* \otimes A \subset \hat{A} \otimes \hat{A}$ with $\sigma : A \otimes V^* \to V^* \otimes A, a \otimes u^* \to u^* \otimes a$, being the switch operator.

Theorem 3.8. *Let A be a k-algebra and let (V, ℓ, r) be an A-bimodule, both with finite k-dimensions. Let $\alpha, \beta : V \to A$ be two k-linear maps. Let $\tilde{\alpha}_-$ and $\tilde{\beta}_+$ be defined by Eq. (25) and identified as linear maps from $\hat{A}^*$ to $\hat{A}$. Then α is an extended $\mathcal{O}$-operator with modification β of mass κ if and only if $\tilde{\alpha}_-$ is an extended $\mathcal{O}$-operator with modification $\tilde{\beta}_+$ of mass κ.*

By Theorem 3.8, the results from the previous sections on $\mathcal{O}$-operators on A can be applied to general $\mathcal{O}$-operators.

Corollary 3.9. *Let A be a k-algebra and let V be an A-bimodule, both with finite k-dimensions.*

(a) *Suppose the characteristic of k is not 2. Let $\alpha, \beta : V \to A$ be linear maps which are identified as elements in $(A \ltimes_{r^*, \ell^*} V^*) \otimes (A \ltimes_{r^*, \ell^*} V^*)$. Then α is an extended $\mathcal{O}$-operator with modification β of mass κ if and only if $(\alpha - \alpha^{21}) \pm (\beta + \beta^{21})$ are solutions of EAYBE of mass $\frac{\kappa+1}{4}$ in $A \ltimes_{r^*, \ell^*} V^*$.*

(b) *Let $\alpha : V \to A$ be a linear map which is identified as an element in $(A \ltimes_{r^*, \ell^*} V^*) \otimes (A \ltimes_{r^*, \ell^*} V^*)$. Then α is an $\mathcal{O}$-operator of weight* *zero if and only if $\alpha - \alpha^{21}$ is a skew-symmetric solution of AYBE in $A \ltimes_{r^*, \ell^*} V^*$. In particular, a linear map $P : A \to A$ is a Rota-Baxter operator of weight zero if and only if $r = P - P^{21}$ is a skew-symmetric solution of AYBE in $A \ltimes_{R^*, L^*} A^*$.*

(c) *Let $\alpha, \beta : V \to A$ be two linear maps which are identified as elements in $(A \ltimes_{r^*, \ell^*} V^*) \otimes (A \ltimes_{r^*, \ell^*} V^*)$. Then α is an extended $\mathcal{O}$-operator with modification β of mass -1 if and only if $(\alpha - \alpha^{21}) \pm (\beta + \beta^{21})$ are solutions of AYBE in $A \ltimes_{r^*, \ell^*} V^*$.*

(d) *Let $\alpha : A \to A$ be a linear map which is identified as an element in $(A \ltimes_{R^*, L^*} A^*) \otimes (A \ltimes_{R^*, L^*} A^*)$. Then α satisfies Eq. (17) if and only if $(\alpha - \alpha^{21}) \pm (\mathrm{id} + \mathrm{id}^{21})$ are solutions of AYBE in $A \ltimes_{R^*, L^*} A^*$.*

(e) *Let $P : A \to A$ be a linear map which is identified as an element of $A \ltimes_{R^*, L^*} A^*$. Then P is a Rota-Baxter operator of weight $\lambda \neq 0$ if and only if both $\frac{2}{\lambda}(P - P^{21}) + 2\mathrm{id}$ and $\frac{2}{\lambda}(P - P^{21}) - 2\mathrm{id}^{21}$ are solutions of AYBE in $A \ltimes_{R^*, L^*} A^*$.*

3.3.2. *The case of Frobenius algebras*

We end this section by considering the relationship between $\mathcal{O}$-operators and solutions of AYBE on Frobenius algebras.

Definition 3.10.

(a) Let A be a **k**-algebra and let $B(\ , \ ) : A \otimes A \to \mathbf{k}$ be a nondegenerate bilinear form. Let $\varphi : A \to A^*$ denote the induced injective linear map defined by

$$B(x, y) = \langle \varphi(x), y \rangle, \quad \forall x, y \in A. \tag{26}$$

(b) A **Frobenius k-algebra** is a **k**-algebra $(A, \cdot)$ together with a nondegenerate bilinear form $B(\ , \ ) : A \otimes A \to \mathbf{k}$ that is **invariant** in the sense that

$$B(x \cdot y, z) = B(x, y \cdot z), \ \forall x, y, z \in A. \tag{27}$$

We use $(A, \cdot, B)$ to denote a Frobenius **k**-algebra.

(c) A Frobenius **k**-algebra is called **symmetric** if

$$B(x, y) = B(y, x), \ \forall x, y \in A.$$

(d) A linear map $\beta : A \to A$ is called **self-adjoint** (resp. **skew-adjoint**) with respect to a bilinear form B if for any $x, y \in A$, we have $B(\beta(x), y) = B(x, \beta(y))$ (resp. $B(\beta(x), y) = -B(x, \beta(y))$).

We now apply results from Section 3.2 to the case of Frobenius algebras.

Theorem 3.11. *Let* **k** *have characteristic not equal to 2. Let* $(A, \cdot, B)$ *be a symmetric Frobenius algebra of finite* **k**-*dimension. Suppose that* α *and* β *are two endomorphisms of* A *and* β *is self-adjoint with respect to* B.

(a) *α is an extended $\mathcal{O}$-operator with modification β of mass κ if and only if $\tilde{\alpha} := \alpha \circ \varphi^{-1} : A^* \to A$ is an extended $\mathcal{O}$-operator with modification $\tilde{\beta} := \beta \circ \varphi^{-1} : A^* \to A$ of mass κ, where the linear map $\varphi : A \to A^*$ is defined by Eq. (26).*

(b) *Suppose that, in addition, α is skew-adjoint with respect to B. Then $\tilde{\alpha}$ regarded as an element of $A \otimes A$ is skew-symmetric and we have*

   (i) *$r_\pm = \tilde{\alpha} \pm \tilde{\beta}$ regarded as elements of $A \otimes A$ are solutions of EAYBE of mass $\frac{\kappa+1}{4}$ if and only if α is an extended $\mathcal{O}$-operator with modification β of mass κ.*

   (ii) *If $\kappa = -1$, then $r_\pm = \tilde{\alpha} \pm \tilde{\beta}$ regarded as elements of $A \otimes A$ are solutions of AYBE if and only if α is an extended $\mathcal{O}$-operator with modification β of mass -1.*

   (iii) *If $\kappa = 0$, then $\tilde{\alpha}$ regarded as an element of $A \otimes A$ is a solution of AYBE if and only if α is a Rota-Baxter operator of weight zero.*

Corollary 3.12. *Let* **k** *have characteristic not equal to 2. Let A be a* **k**-*algebra of finite* **k**-*dimension and let $r \in A \otimes A$. Define $\alpha, \beta \in A \otimes A$ by Eq. (22). Let $B : A \otimes A \to \mathbf{k}$ be a nondegenerate symmetric and invariant bilinear form. Define the linear map $\varphi : A \to A^*$ by Eq. (26).*

(a) *Suppose that $\beta \in A \otimes A$ is invariant. Then r is a solution of EAYBE of mass $\frac{\kappa+1}{4}$ if and only if $\hat{\alpha} = \alpha\varphi : A \to A$ is an extended $\mathcal{O}$-operator with modification $\hat{\beta} = \beta\varphi : A \to A$ of mass k.*

(b) *Suppose that $\beta \in A \otimes A$ is invariant. Then r is a solution of AYBE if and only if $\hat{\alpha} = \alpha\varphi : A \to A$ is an extended $\mathcal{O}$-operator with modification $\hat{\beta} = \beta\varphi : A \to A$ of mass -1. In particularly, if in addition, $\beta = 0$, i.e., r is skew-symmetric, then r is a solution of AYBE if and only if $\hat{\alpha} = \hat{r} = r\varphi : A \to A$ is a Rota-Baxter operator of weight zero.*

4. Antisymmetric infinitesimal bialgebras and generalized AYBE

In Section 4.1, we define an antisymmetric infinitesimal bialgebra and discuss its close tie with generalized AYBE. We then consider the relationship between extended $\mathcal{O}$-operators and the generalized AYBE. The factorization problem of an antisymmetric infinitesimal bialgebra and the double space of an factorizable antisymmetric infinitesimal bialgebra are studied in Section 4.2.

4.1. Antisymmetric infinitesimal bialgebras, generalized AYBE and extended $\mathcal{O}$-operators

We start with the definition of an antisymmetric infinitesimal bialgebra.

Definition 4.1. ([6]) Let A be a k-algebra. An **antisymmetric infinitesimal bialgebra** structure on A is a coproduct $\Delta : A \to A \otimes A$ such that

(a) (A, Δ) is a coassociative coalgebra, i.e., $\Delta^* : A^* \otimes A^* \to A^*$ induces a k-algebra structure on A^*;

(b) the following two equations hold:

$$\Delta(x \circ y) = (\mathrm{id} \otimes L(x))\Delta(y) + (R(y) \otimes \mathrm{id})\Delta(x), \qquad (28)$$

$$(L(y)\otimes\mathrm{id}-\mathrm{id}\otimes R(y))\Delta(x)+\sigma[(L(x)\otimes\mathrm{id}-\mathrm{id}\otimes R(x))\Delta(y)] = 0, \ \forall x, y \in A. \qquad (29)$$

It is denoted by (A, Δ). An antisymmetric infinitesimal bialgebra (A, Δ) is called **coboundary** if there exists an $r \in A \otimes A$ such that

$$\Delta(x) = (\mathrm{id} \otimes L(x) - R(x) \otimes \mathrm{id})r, \quad \forall x \in A. \qquad (30)$$

It is usually denoted by (A, r).

In fact, an antisymmetric infinitesimal bialgebra is exactly an associative D-bialgebra[34] and a balanced infinitesimal bialgebra in the opposite algebra.[3]

Definition 4.2. An element $r \in A \otimes A$ is called a solution of the **generalized associative Yang-Baxter equation (GAYBE) in** A if it satisfies the relation

$$(\mathrm{id}\otimes\mathrm{id}\otimes L(x) - R(x)\otimes\mathrm{id}\otimes\mathrm{id})(r_{12}r_{13}+r_{13}r_{23}-r_{23}r_{12}) = 0, \quad \forall x \in A. \quad (31)$$

The following results demonstrate the close relationship between GAYBE and an antisymmetric infinitesimal bialgebra.

Proposition 4.3. ([1,3,6]) Let A be a k-algebra and let $r \in A \otimes A$. Define a coproduct $\Delta : A \to A \otimes A$ by Eq. (30). Then (A, Δ) is a coassociative coalgebra (*i.e. Item (a) in Definition 4.1 holds*) *if and only if* r *is a solution of GAYBE.*

Theorem 4.4. (^{6}) *Let* A *be a* **k**-*algebra and* $r \in A \otimes A$. *Then the co-operation defined by Eq. (30) satisfies Eq. (28) and Eq. (29) if and only if*

$$(L(x)\otimes\mathrm{id}-\mathrm{id}\otimes R(x))(\mathrm{id}\otimes L(y)-R(y)\otimes\mathrm{id})(r+\sigma(r)) = 0, \quad \forall x, y \in A. \quad (32)$$

Therefore (A, r) *becomes a coboundary antisymmetric infinitesimal bialgebra if and only if* r *is a solution of GAYBE satisfying Eq. (32).*

We now consider the relationship between extended $\mathcal{O}$-operators, GAYBE and antisymmetric infinitesimal bialgebras. We first start with a 2-tensor $r \in A^{\otimes 2}$ and show how r being a solution of GAYBE or giving rise to a coboundary infinitesimal bialgebra can be determined by its associated linear operators being an extended $\mathcal{O}$-operator.

Proposition 4.5. *Let* **k** *have characteristic not equal to 2. Let* $(A, \cdot)$ *be a* **k**-*algebra with finite* **k**-*dimension and* $r \in A \otimes A$. *Let* α, β *be given by Eq. (22). Suppose that* β *is a balanced* A-*bimodule homomorphism, that is, β satisfies Eq. (23).*

(a) *If* α *is an extended* $\mathcal{O}$-*operator with modification* β *of mass* $\kappa \in$ **k**, *then* r *is a solution of GAYBE.*

(b) *Under the same conditions as in Item (a),* (A, r) *is a coboundary antisymmetric infinitesimal bialgebra.*

(c) *If* r *is a solution of MAYBE of mass* $\kappa \in$ **k** *and its symmetric part is invariant, then it is also a solution of GAYBE.*

We next start with an extended $\mathcal{O}$-operator and characterize it in terms of the associated 2-tensor being a solution of GAYBE.

Proposition 4.6. *Let* $(A, \cdot)$ *be a* **k**-*algebra with finite* **k**-*dimension.*

(a) *Let* $(R, \circ, \ell, r)$ *be an* A-*bimodule* **k**-*algebra with finite* **k**-*dimension. Let* $\alpha, \beta : R \to A$ *be two linear maps such that* α *is an extended* $\mathcal{O}$-*operator of weight* λ *with modification* β *of mass* (κ, μ). *Then* $\alpha - \alpha^{21}$ *identified as an element of* $(A \ltimes_{r^*, \ell^*} R^*) \otimes (A \ltimes_{r^*, \ell^*} R^*)$ *is a skew-symmetric solution of GAYBE (31) if and only if the following equations hold:*

$$\lambda\ell(\alpha(u \circ v))w = \lambda ur(\alpha(v \circ w)), \quad \forall u, v, w \in R, \quad (33)$$

$$\lambda\alpha(u \circ (vr(x))) = \lambda\alpha(u \circ v) \cdot x, \quad \forall u, v \in R, x \in A, \quad (34)$$

$$\lambda\alpha((\ell(x)u) \circ v) = \lambda x \cdot \alpha(u \circ v), \quad \forall u, v \in R, x \in A. \tag{35}$$

In particular, when $\lambda = 0$, *then* $\alpha - \alpha^{21}$ *identified as an element of* $(A \ltimes_{r^*,\ell^*} R^*) \otimes (A \ltimes_{r^*,\ell^*} {}^`R^*)$ *is a skew-symmetric solution of GAYBE (31).*

(b) *Let* $(R, \circ, \ell, r)$ *be an* A-*bimodule* **k**-*algebra with finite* **k**-*dimension. Let* $\alpha : R \to A$ *be an* $\mathcal{O}$-*operator of weight* λ. *Then* $\alpha - \alpha^{21}$ *identified as an element of* $(A \ltimes_{r^*,\ell^*} R^*) \otimes (A \ltimes_{r^*,\ell^*} R^*)$ *is a skew-symmetric solution of GAYBE if and only if Eq. (33), Eq. (34) and Eq. (35) hold.*

(c) *Let* (V, ℓ, r) *be an* A-*bimodule with finite* **k**-*dimension. Let* $\alpha, \beta : V \to A$ *be two linear maps such that* α *is an extended* $\mathcal{O}$-*operator with modification* β *of mass* κ. *Then* $\alpha - \alpha^{21}$ *identified as an element of* $(A \ltimes_{r^*,\ell^*} V^*) \otimes (A \ltimes_{r^*,\ell^*} V^*)$ *is a skew-symmetric solution of GAYBE.*

(d) *Let* $\alpha : A \to A$ *be a linear endomorphism of* A. *Suppose that* α *satisfies Eq. (16). Then* $\alpha - \alpha^{21}$ *identified as an element of* $(A \ltimes_{R^*,L^*} A^*) \otimes (A \ltimes_{R^*,L^*} A^*)$ *is a skew-symmetric solution of GAYBE.*

(e) *Let* $(R, \circ, \ell, r)$ *be an* A-*bimodule* **k**-*algebra of finite* **k**-*dimension. Let* $\alpha, \beta : R \to A$ *be two linear maps such that* α *is an extended* $\mathcal{O}$-*operator with modification* β *of mass* $(\kappa, \mu) = (0, \mu)$, *i.e.,* β *is an* A-*bimodule homomorphism and the condition (11) in Definition 2.5 holds, and* α *and* β *satisfy the following equation:*

$$\alpha(u) \cdot \alpha(v) - \alpha(\ell(\alpha(u))v + ur(\alpha(v))) = \mu\beta(u \circ v), \quad \forall u, v \in R.$$

Then $\alpha - \alpha^{21}$ *identified as an element of* $(A \ltimes_{r^*,\ell^*} R^*) \otimes (A \ltimes_{r^*,\ell^*} R^*)$ *is a skew-symmetric solution of GAYBE.*

4.2. *Factorizable quasitriangular antisymmetric infinitesimal bialgebras*

This subsection is devoted to study the so-called **factorizable quasitriangular antisymmetric infinitesimal bialgebras** and their relationship with the factorization problem.

Definition 4.7. Let (A, r) be a coboundary antisymmetric infinitesimal bialgebra.

(a) $(^3)$ (A, r) is called **quasitriangular** if r is a solution of AYBE.

(b) $(^3)$ (A, r) is called **triangular** if r is a skew-symmetric solution of AYBE.

(c) Suppose (A, r) is quasitriangular. (A, r) is called **factorizable** if $r + \sigma(r) = r + r^t$ regarded as a linear map from A^* to A is invertible.

Let (A, r) be a factorizable quasitriangular antisymmetric infinitesimal bialgebra. Since $r + r^{21}$ is invertible, we obtain the following conclusion, which justifies the term factorizable. In the Lie algebra context, the factorizable property is important in integrable systems.[29,32]

Proposition 4.8. *Let (A, r) be a factorizable quasitriangular antisymmetric infinitesimal bialgebra with finite $\mathbf{k}$-dimension. Then any element $x \in A$ admits a unique decomposition*

$$x = x_+ + x_- \tag{36}$$

with $(x_+, x_-) \in \mathrm{Im}(r \oplus r^t) \subset A \oplus A$.

Proof. Since $r + r^t = 2\beta : A^* \to A$ is invertible, for $x \in A$ we have

$$x = r\left(\frac{\beta^{-1}(x)}{2}\right) + r^t\left(\frac{\beta^{-1}(x)}{2}\right) \in \mathrm{Im}(r \oplus r^t) \subset A \oplus A.$$

On the other hand, if there exist $a^*, b^* \in A^*$ such that $x = r(a^*) + r^t(a^*) = r(b^*) + r^t(b^*)$. Then $0 = r(a^* - b^*) + r^t(a^* - b^*) = 2\beta(a^* - b^*)$. Since $\beta : A^* \to A$ is invertible, we obtain $a^* = b^*$. So the conclusion follows. $\square$

Proposition 4.9. ([6]) *Let (A, Δ) be an antisymmetric infinitesimal bialgebra with A being of finite $\mathbf{k}$-dimension.*

(a) *There exists a unique $\mathbf{k}$-algebra structure on $\mathcal{AD}(A) := A \oplus A^*$ such that A and A^* are $\mathbf{k}$-subalgebras and the natural symmetric (nondegenerate) bilinear form on $\mathcal{AD}(A)$*

$$B_p((x, a^*), (y, b^*)) = \langle a^*, y \rangle + \langle x, b^* \rangle, \quad \forall x, y \in A; a^*, b^* \in A^*,$$

is invariant in the sense of Eq. (27). Explicitly, the $\mathbf{k}$-algebra structure on $\mathcal{AD}(A)$ is defined by

$$(x, a^*) \cdot (y, b^*) = (x \circ y + R^*(a^*)y + xL^*(b^*), a^* * b^* + R^*(x)b^* + a^*L^*(y)), \tag{37}$$

for all $x, y \in A, a^*, b^* \in A^*$. Here the $\mathbf{k}$-algebra structure $*$ on A^* is given by $\Delta^* : A^* \otimes A^* \cong (A \otimes A)^* \to A^*$.

(b) *There exist a natural coboundary infinitesimal bialgebra structure on $\mathcal{AD}(A)$ which is induced by $r = \sum_i e_i \otimes e_i^*$, where $\{e_i\}_{1 \leq i \leq \dim A}$ is a basis of A and $\{e_i^*\}_{1 \leq i \leq \dim A}$ is the dual basis.*

The associative algebra $\mathcal{AD}(A) = A \oplus A^*$ in Proposition 4.9 is called its **double space**.

Lemma 4.10. $(^9)$ *Let* $(A, \cdot)$ *be a* **k**-*algebra with finite* **k**-*dimension. Let* $r \in A \otimes A$. *Define a coproduct* $\Delta : A \to A \otimes A$ *by Eq. (30). Then the multiplication* $*$ *on* A^* *given by* $\Delta^* : A^* \otimes A^* \cong (A \otimes A)^* \to A^*$ *is*

$$a^* * b^* = R^*(r(a^*))b^* - L^*(r^t(b^*))a^*, \quad \forall a^*, b^* \in A^*. \tag{38}$$

Proposition 4.11. *Let* (A, r) *be a factorizable quasitriangular antisymmetric infinitesimal bialgebra with finite* **k**-*dimension. Then its double space* $\mathcal{AD}(A)$ *is isomorphic to the direct sum* $A \oplus A$ *of* **k**-*algebras.*

Proof. We prove that the linear map $\theta : \mathcal{AD}(A) = A \oplus A^* \to A \oplus A$ defined by

$$\theta(x) = (x, x), \quad \theta(a^*) = (r(a^*), -r^t(a^*)), \quad \forall x \in A, a^* \in A^*,$$

is an isomorphism of **k**-algebras. It is obvious that $\theta|_A$ is an embedding of **k**-algebras. Define α and β by Eq. (22). Then by Corollary 3.5, Eq. (24) holds. So $\theta|_{A^*}$ is a homomorphism of **k**-algebras. Moreover, for any $x \in A, a^*, b^* \in A^*$,

$$
\begin{aligned}
\langle xL^*(a^*), b^* \rangle &= \langle x, a^* * b^* \rangle = \langle x, R^*(r(a^*))b^* - a^*L^*(r^t(b^*)) \rangle \\
&= \langle x, R^*(\alpha(a^*))b^* + a^*L^*(\alpha(b^*)) \rangle \text{ (by the invariance of } \beta) \\
&= \langle x \circ \alpha(a^*) - \alpha(R^*(x)a^*), b^* \rangle \text{ (by the skew} - \text{symmetry of } \alpha).
\end{aligned}
$$

Thus,

$$xL^*(a^*) = x \circ \alpha(a^*) - \alpha(R^*(x)a^*), \quad \forall x \in A, a^* \in A^*. \tag{39}$$

Therefore,

$$
\begin{aligned}
\theta(x \cdot a^*) &= \theta(R^*(x)a^* + xL^*(a^*)) = ((\alpha + \beta)(R^*(x)a^*), (\alpha - \beta)(R^*(x)a^*)) \\
&\quad + (xL^*(a^*), xL^*(a^*)) \\
&= (\alpha(R^*(x)a^*) + x \circ \beta(a^*) + xL^*(a^*), \alpha(R^*(x)a^*) - x \circ \beta(a^*) + xL^*(a^*)) \\
&\quad \text{(by the invariance of } \beta) \\
&= (x \circ (\alpha + \beta)(a^*), x \circ (\alpha - \beta)(a^*)) \quad \text{(by Eq. (39))} \\
&= (x, x) \circ ((\alpha + \beta)(a^*), (\alpha - \beta)(b^*)) \\
&= \theta(x) \cdot \theta(a^*).
\end{aligned}
$$

We similarly prove that

$$\theta(a^* \cdot x) = \theta(a^*) \cdot \theta(x), \quad \forall x \in A, a^* \in A^*.$$

On the other hand, since β is invertible, θ is bijective. So θ is an isomorphism of **k**-algebras. $\square$

5. $\mathcal{O}$-operators and dendriform algebras

In this section we study the relationship of $\mathcal{O}$-operators with dendriform dialgebras and trialgebras. This section is based on[10] where the interested reader can find further details.

Given an A-bimodule V and an $\mathcal{O}$-operator $\alpha : V \to A$, as in the case of a Rota-Baxter operator, we obtain a dendriform dialgebra $(V, \prec_\alpha, \succ_\alpha)$. We also show that an $\mathcal{O}$-operator on an algebra gives a dendriform trialgebra. We prove in Section 5.2 that every dendriform dialgebras and trialgebra can be recovered from an $\mathcal{O}$-operator in this way, in contrary to the case of a Rota-Baxter operator.

In Section 5.3 we further show that the dendriform dialgebra or trialgebra structure on V from an $\mathcal{O}$-operator $\alpha : V \to A$ transports to a dendriform dialgebra or trialgebra structure on A through α under a natural condition. To distinguish the two dendriform dialgebras and trialgebras from an $\mathcal{O}$-operator $\alpha : V \to A$, we call them the **dendriform dialgebras and trialgebras on the domain** and the **dendriform dialgebras and trialgebras on the range** of α respectively.

By considering the multiplication on the range A, we show that, the correspondence from $\mathcal{O}$-operators to dendriform dialgebras and trialgebras on the domain V implies a more refined correspondence from $\mathcal{O}$-operators to dendriform dialgebras and trialgebra on the range A that are compatible with A in the sense that the dialgebra and trialgebra multiplications give a splitting (or decomposition) of the associative product of A. We finally quantify this refined correspondence by providing bijections between certain equivalent classes of $\mathcal{O}$-operators with range in A and equivalent classes of compatible dendriform dialgebra and trialgebra structures on A.

5.1. *Rota-Baxter algebras and dendriform algebras*

Generalizing the concept of a dendriform dialgebra of Loday defined in Section 1, the concept of a dendriform trialgebra was introduced by Loday and Ronco.

Definition 5.1. ([28]) Let **k** be a commutative ring. A **dendriform k-trialgebra** is a quadruple $(T, \prec, \succ, \cdot)$ consisting of a **k**-vector space T and three binary operations $\prec$, $\succ$ and $\cdot$ such that

$$(x \prec y) \prec z = x \prec (y \star z), \quad (x \succ y) \prec z = x \succ (y \prec z),$$

$$(x \star y) \succ z = x \succ (y \succ z), \quad (x \succ y) \cdot z = x \succ (y \cdot z),$$

$$(x \prec y) \cdot z = x \cdot (y \succ z), \quad (x \cdot y) \prec z = x \cdot (y \prec z), \quad (x \cdot y) \cdot z = x \cdot (y \cdot z)$$

for all $x, y, z \in T$. Here $\star = \prec + \succ + \cdot$.

It is easy to check[27],[28] that, given a dendriform dialgebra $(D, \prec, \succ)$ (resp. dendriform trialgebra $(D, \prec, \succ, \cdot)$), the product given by

$$x \star y = x \prec y + x \succ y, \quad \forall x, y \in D$$

(resp.

$$x \star y = x \prec y + x \succ y + x \cdot y, \quad \forall x, y \in D)$$

defines a **k**-algebra product on D. We say that dendriform dialgebra (resp. trialgebra) gives a **splitting** of the associative multiplication $\star$.

Generalizing the result of Aguiar in Eq. (6), Ebrahimi-Fard[16] showed that, if $(R, \circ, P)$ is a Rota-Baxter algebra of weight $\lambda \neq 0$, then the multiplications

$$x \prec_P y := x \circ P(y), \quad x \succ_P y := P(x) \circ y, \quad x \cdot_P y := \lambda x \circ y, \quad \forall x, y \in R, \tag{40}$$

define a dendriform trialgebra $(R, \prec_P, \succ_P, \cdot_P)$.

For a given **k**-vector space V, define

$$\mathbf{RB}_\lambda(V) := \left\{ (V, \circ, P) \middle| \begin{array}{l} (V, \circ) \text{ is an } \mathbf{k}\text{-algebra and} \\ P \text{ is a Rota-Baxter operator of weight } \lambda \text{ on } (V, \circ) \end{array} \right\} \tag{41}$$

$$\mathbf{DD}(V) := \{(V, \prec, \succ) \mid (V, \prec, \succ) \text{ is a dendriform dialgebra}\}, \tag{42}$$

$$\mathbf{DT}(V) := \{(V, \prec, \succ, \cdot) \mid (V, \prec, \succ, \cdot) \text{ is a dendriform trialgebra}\}. \tag{43}$$

Then Eq. (40) yields a map

$$\Phi_{V,\lambda} : \mathbf{RB}_\lambda(V) \longrightarrow \mathbf{DT}(V)$$

which, when $\lambda = 0$, reduces to the map

$$\Phi_{V,0} : \mathbf{RB}_0(V) \longrightarrow \mathbf{DD}(V)$$

from Eq. (6). Thus deriving all dendriform dialgebras (resp. trialgebras) on V from Rota-Baxter operators on V amounts to the surjectivity of $\Phi_{V,0}$ (resp. $\Phi_{V,\lambda}$).

Unfortunately this map is quite far away from being surjective. See the examples in.[10]

5.2. *From $\mathcal{O}$-operators to dendriform algebras on the domains*

We first generalize the correspondence from Rota-Baxter operators to dendriform algebras to that from $\mathcal{O}$-operators.

Theorem 5.2. *Let $(A, \cdot)$ be a **k**-algebra.*

(a) *Let $(R, \circ, \ell, r)$ be an A-bimodule $\mathbf{k}$-algebra. Let $\alpha : R \to A$ be an $\mathcal{O}$-operator on the algebra R of weight λ. Then the multiplications*

$$u \prec_\alpha v := ur(\alpha(v)), \quad u \succ_\alpha v := \ell(\alpha(u))v, \quad u \cdot_\alpha v := \lambda\, u \circ v, \quad \forall u, v \in R,$$

define a dendriform trialgebra $(R, \prec_\alpha, \succ_\alpha, \cdot_\alpha)$. Further, the multiplication $\star_\alpha := \prec_\alpha + \succ_\alpha + \cdot_\alpha$ on R defines a $\mathbf{k}$-algebra product on R and the map $\alpha : (R, \star_\alpha) \to (A, \cdot)$ is a $\mathbf{k}$-algebra homomorphism.

(b) *Let (V, ℓ, r) be an A-bimodule. Let $\alpha : V \to A$ be an $\mathcal{O}$-operator on the vector space V. Then the multiplications*

$$u \prec_\alpha v := ur(\alpha(v)), \quad u \succ_\alpha v := \ell(\alpha(u))v, \quad \forall u, v \in V,$$

define a dendriform dialgebra $(V, \prec_\alpha, \succ_\alpha)$. Further, the multiplication $\star_\alpha := \prec_\alpha + \succ_\alpha$ on V defines a $\mathbf{k}$-algebra product on V and $\alpha : (V, \star_\alpha) \to (A, \cdot)$ is a $\mathbf{k}$-algebra homomorphism.

We call the dendriform trialgebra (resp. dialgebra) obtained in the theorem the **dendriform trialgebra (resp. dialgebra) on the domain of** α to distinguish it from a similar structure to be introduced in Proposition 5.4. Theorem 5.2.(b) was obtained independently in.[33]

For a $\mathbf{k}$-algebra A and an A-bimodule $\mathbf{k}$-algebra $(R, \circ)$, denote

$$\mathbf{O}_\lambda^{\mathrm{alg}}(R, A) := \mathbf{O}_\lambda^{\mathrm{alg}}((R, \circ), A) := \left\{ \alpha : R \to A \,\middle|\, \begin{array}{l} \alpha \text{ is an } \mathcal{O}\text{-operator} \\ \text{of weight } \lambda \text{ on the algebra } R \end{array} \right\}.$$

By Theorem 5.2.(a), we obtain a map

$$\Phi_{\lambda, R, A}^{\mathrm{alg}} : \mathbf{O}_\lambda^{\mathrm{alg}}((R, \circ), A) \longrightarrow \mathbf{DT}(|R|), \tag{44}$$

where $|R|$ denotes the underlying vector space of R.

Now let V be a $\mathbf{k}$-vector space. Let $\mathbf{O}_\lambda^{\mathrm{alg}}(V, -)$ denote the set of $\mathcal{O}$-operators on the algebra $(V, \circ)$ of weight λ, where $\circ$ is a $\mathbf{k}$-algebra product on V. In other words,

$$\mathbf{O}_\lambda^{\mathrm{alg}}(V, -) := \coprod_{R, A} \mathbf{O}_\lambda^{\mathrm{alg}}(R, A),$$

where the disjoint union runs through all pairs (R, A) where A is a $\mathbf{k}$-algebra and R is an A-bimodule $\mathbf{k}$-algebra such that $|R| = V$. Then from the map $\Phi_{\lambda, V, A}^{\mathrm{alg}}$ in Eq. (44) we obtain

$$\Phi_{\lambda, V}^{\mathrm{alg}} := \coprod_{R, A} \Phi_{\lambda, V, A}^{\mathrm{alg}} : \mathbf{O}_\lambda^{\mathrm{alg}}(V, -) \longrightarrow \mathbf{DT}(V).$$

Similarly, for a **k**-vector space V and **k**-algebra A, denote

$$\mathbf{O}^{\mathrm{mod}}(V, A) = \{\alpha : V \to A \mid \alpha \text{ is an } \mathcal{O}\text{-operator on the vector space } V\}.$$

By Theorem 5.2.(b), we obtain a map

$$\Phi_{V,A}^{\mathrm{mod}} : \mathbf{O}^{\mathrm{mod}}(V, A) \longrightarrow \mathbf{DD}(V).$$

Let $\mathbf{O}^{\mathrm{mod}}(V, -)$ denote the set of $\mathcal{O}$-operators on the vector space V. In other words,

$$\mathbf{O}^{\mathrm{mod}}(V, -) := \coprod_A \mathbf{O}^{\mathrm{mod}}(V, A),$$

where A runs through all the **k**-algebras. Then we obtain a map

$$\Phi_V^{\mathrm{mod}} := \coprod_A \Phi_{V,A}^{\mathrm{mod}} : \mathbf{O}^{\mathrm{mod}}(V, -) \longrightarrow \mathbf{DD}(V).$$

Let us compare $\Phi_{0,V}^{\mathrm{alg}}$ and Φ_V^{mod} for a **k**-vector space V. For a given **k**-algebra multiplication $\circ$ on V, we have the natural bijection $\mathbf{O}_0^{\mathrm{alg}}((V, \circ), -) \to \mathbf{O}^{\mathrm{mod}}(V, -)$ sending an $\mathcal{O}$-operator $\alpha : (V, \circ) \to A$ on the algebra $(V, \circ)$ to the $\mathcal{O}$-operator $\alpha : V \to A$ on the underlying **k**-vector space V. Thus $\mathbf{O}_0^{\mathrm{alg}}(V, -)$ is the disjoint union of multiple copies of $\mathbf{O}^{\mathrm{mod}}(V, -)$, one copy for each **k**-algebra multiplication on V. Therefore, the surjectivity of Φ_V^{mod} is a stronger property than the surjectivity of $\Phi_{0,V}^{\mathrm{alg}}$.

Theorem 5.3. *Let V be a **k**-vector space. The maps $\Phi_{1,V}^{\mathrm{alg}}$ and Φ_V^{mod} are surjective.*

By this theorem, all dendriform dialgebra (resp. trialgebra) structures on V are recovered from $\mathcal{O}$-operators on the vector space (resp. on the algebra).

5.3. *From $\mathcal{O}$-operators to dendriform algebras on the ranges*

We now consider the dendriform algebra structures on the range of an $\mathcal{O}$-operator.

Proposition 5.4. *Let $(A, \cdot)$ be a **k**-algebra.*

*(a) Let $(R, \circ, \ell, r)$ be an A-bimodule **k**-algebra. Let $\alpha : R \to A$ be an $\mathcal{O}$-operator on the algebra of weight λ. If $\ker \alpha$ is an ideal of $(R, \circ)$, then there is a dendriform trialgebra structure on $\alpha(R)$ given by*

$$\alpha(u) \prec_{\alpha, A} \alpha(v) := \alpha(ur(\alpha(v))), \quad \alpha(u) \succ_{\alpha, A} \alpha(v) := \alpha(\ell(\alpha(u))v),$$
$$\alpha(u) \cdot_{\alpha, A} \alpha(v) := \alpha(\lambda u \circ v), \quad \forall u, v \in R.$$

Furthermore, $\cdot = \prec_{\alpha,A} + \succ_{\alpha,A} + \cdot_{\alpha,A}$ *on* $\alpha(R)$. *In particular, if the* $\mathcal{O}$-*operator* α *is invertible (that is, bijective as a* **k**-*linear map), then the multiplications*

$$x \prec_{\alpha,A} y := \alpha(\alpha^{-1}(x)r(y)), \quad x \succ_{\alpha,A} y := \alpha(\ell(x)\alpha^{-1}(y)),$$
$$x \cdot_{\alpha,A} y := \alpha(\lambda\alpha^{-1}(x) \circ \alpha^{-1}(y)), \quad \forall x,y \in A,$$

define a dendriform trialgebra $(A, \prec_{\alpha,A}, \succ_{\alpha,A}, \cdot_{\alpha,A})$ *such that* $\cdot = \prec_{\alpha,A} + \succ_{\alpha,A} + \cdot_{\alpha,A}$ *on* A, *called the* **dendriform trialgebra on the range of** α.

(b) *Let* (V, ℓ, r) *be a* A-*bimodule. Let* $\alpha : V \to A$ *be an invertible* $\mathcal{O}$-*operator on the vector space. Then*

$$x \prec_{\alpha,A} y := \alpha(\alpha^{-1}(x)r(y)), \quad x \succ_{\alpha,A} y := \alpha(\ell(x)\alpha^{-1}(y)), \quad \forall x,y \in A,$$

define a dendriform dialgebra $(A, \prec_{\alpha,A}, \succ_{\alpha,A})$ *on* A *such that* $\cdot = \prec_{\alpha,A} + \succ_{\alpha,A}$ *on* A, *called the* **dendriform dialgebra on the range** of α.

Proposition 5.4 motivates us to introduce the following notations.

Definition 5.5. Let $(A, \cdot)$ be a **k**-algebra and let $\lambda \in$ **k**.

(a) Let $\mathbf{IO}_\lambda^{\mathrm{alg}}(A, \cdot)$ (resp. $\mathbf{IO}^{\mathrm{mod}}(A, \cdot)$) denote the set of invertible (i.e., bijective) $\mathcal{O}$-operators $\alpha : R \to A$ on the algebra of weight $\lambda \in$ **k** (resp. on the vector space), where $R = (R, \circ, \ell, r)$ is an A-bimodule **k**-algebra (resp. $R = (R, \ell, r)$ is an A-bimodule).

(b) Let $\mathbf{DT}(A, \cdot)$ (resp. $\mathbf{DD}(A, \cdot)$) denote the set of dendriform trialgebra (resp. dialgebra) structures $(A, \prec, \succ, \cdot)$ (resp. $(A, \prec, \succ)$) on $(A, \cdot)$ such that $\cdot = \prec + \succ + \cdot$ (resp. $\cdot = \prec + \succ$).

(c) Let

$$\Psi_A^{\mathrm{alg}} : \mathbf{IO}_\lambda^{\mathrm{alg}}(A, \cdot) \longrightarrow \mathbf{DT}(A, \cdot), \quad \alpha \mapsto (A, \prec_{\alpha,A}, \succ_{\alpha,A}, \cdot_{\alpha,A}) \quad (45)$$

$$(resp. \quad \Psi_A^{\mathrm{mod}} : \mathbf{IO}^{\mathrm{mod}}(A, \cdot) \longrightarrow \mathbf{DD}(A, \cdot), \quad \alpha \mapsto (A, \prec_{\alpha,A}, \succ_{\alpha,A})) \tag{46}$$

be the map defined by Proposition 5.4.

Instead of proving just the surjectivities of the maps Ψ_A^{alg} and Ψ_A^{mod} defined by Eq. (45) and Eq. (46), we give a more quantitative description of these maps.

Proposition 5.6. *Let* $(A, \cdot)$ *be a* **k**-*algebra and let* $(R, \circ, \ell, r)$ *be an* A-*bimodule* **k**-*algebra. Let* $\alpha : (R, \circ, \ell, r) \to A$ *be an* $\mathcal{O}$-*operator on the algebra* $(R, \circ)$ *of weight* λ.

(a) Let $g : (R_1, \circ_1, \ell_1, r_1) \to (R, \circ, \ell, r)$ be an isomorphism of A-bimodule **k**-algebras. Then $\alpha g : (R_1, \circ_1, \ell_1, r_1) \to A$ is an $\mathcal{O}$-operator on the algebra $(R_1, \circ_1)$ of weight λ.

(b) Let $f : A \to A$ be a **k**-algebra automorphism. Then $f\alpha : (R, \circ, \ell f^{-1}, r f^{-1}) \to A$ is an $\mathcal{O}$-operator on the algebra $(R, \circ)$ of weight λ.

Similar statements hold for an A-bimodule V in place of an A-bimodule **k**-*algebra R.*

Definition 5.7. Let $(A, \cdot)$ be a **k**-algebra and let $\lambda \in \mathbf{k}$.

(a) For A-bimodule **k**-algebras $(R_i, \circ_i, \ell_i, r_i)$ and invertible $\mathcal{O}$-operators $\alpha_i :$ $R_i \to A$ of weight λ, $i = 1, 2$, call α_1 and α_2 **isomorphic**, denoted by $\alpha_1 \cong \alpha_2$, if there is an isomorphism $g : (R_1, \circ_1, \ell_1, r_1) \to (R_2, \circ_2, \ell_2, r_2)$ of A-bimodule **k**-algebras (see Definition 2.2) such that $\alpha_1 = \alpha_2 g$. Similarly define isomorphic invertible $\mathcal{O}$-operators on vector spaces.

(b) For A-bimodule **k**-algebras $(R_i, \circ_i, \ell_i, r_i)$ and invertible $\mathcal{O}$-operators $\alpha_i : R_i \to A$ of weight λ, $i = 1, 2$, call α_1 and α_2 **equivalent**, denoted by $\alpha_1 \sim \alpha_2$, if there exists a **k**-algebra automorphism $f : A \to A$ such that $f\alpha_1 \cong \alpha_2$. In other words, if there exist a **k**-algebra automorphism $f : A \to A$ and an isomorphism $g : (R_1, \circ_1, \ell_1 f^{-1}, r_1 f^{-1}) \to (R_2, \circ_2, \ell_2, r_2)$ of A-bimodule **k**-algebras such that $f\alpha_1 = \alpha_2 g$. Similarly define equivalent invertible $\mathcal{O}$-operators on vector spaces.

(c) Let $\mathbf{IO}_\lambda^{\mathrm{alg}}(A, \cdot)/\cong$ (resp. $\mathbf{IO}_\lambda^{\mathrm{alg}}(A, \cdot)/\sim$) denote the set of equivalent classes from the relation $\cong$ (resp. $\sim$). Similarly define $\mathbf{IO}^{\mathrm{mod}}(A, \cdot)/\cong$ and $\mathbf{IO}^{\mathrm{mod}}(A, \cdot)/\sim$.

(d) Two dendriform trialgebras $(A, \prec_i, \succ_i, \cdot_i), i = 1, 2$, on A are called **isomorphic**, denoted by $(A, \prec_1, \succ_1, \cdot_1) \cong (A, \prec_2, \succ_2, \cdot_2)$ if there is a linear bijection $F : A \to A$ such that

$$F(x \prec_1 y) = F(x) \prec_2 F(y), \ F(x \succ_1 y) = F(x) \succ_2 F(y),$$
$$F(x \cdot_1 y) = F(x) \cdot_2 F(y),$$

for all $x, y \in A$.

(e) Two dendriform dialgebras $(A, \prec_i, \succ_i), i = 1, 2$, on A are called **isomorphic**, denoted by $(A, \prec_1, \succ_1) \cong (A, \prec_2, \succ_2)$ if there is a linear bijection $F : A \to A$ such that

$$F(x \prec_1 y) = F(x) \prec_2 F(y), \quad F(x \succ_1 y) = F(x) \succ_2 F(y), \quad \forall x, y \in A.$$

(f) Let $\mathbf{DT}(A, \cdot)/\cong$ (resp. $\mathbf{DD}(A, \cdot)/\cong$) denote the set of equivalent classes of $\mathbf{DT}(A, \cdot)$ (resp. $\mathbf{DD}(A, \cdot)$) modulo the isomorphisms.

Theorem 5.8. *Let $(A, \cdot)$ be a **k**-algebra and let $\lambda \in \mathbf{k}$. Let*

$$\Psi_A^{\mathrm{alg}} : \mathbf{IO}_\lambda^{\mathrm{alg}}(A, \cdot) \longrightarrow \mathbf{DT}(A, \cdot), \quad \alpha \mapsto (A, \prec_{\alpha,A}, \succ_{\alpha,A}, \cdot_{\alpha,A}),$$

be the map defined by Eq. (45). Then Ψ_A^{alg} induces bijections

$$\Psi_{A,\cong}^{\mathrm{alg}} : \mathbf{IO}_\lambda^{\mathrm{alg}}(A, \cdot)/\cong \; \longrightarrow \mathbf{DT}(A, \cdot),$$

$$\Psi_{A,\sim}^{\mathrm{alg}} : \mathbf{IO}_\lambda^{\mathrm{alg}}(A, \cdot)/\sim \; \longrightarrow \mathbf{DT}(A, \cdot)/\cong.$$

In particular, Ψ_A^{alg} is surjective.

Similar statements hold for Ψ_A^{mod}.

6. $\mathcal{O}$-operators, Rota-Baxter operators, relative differential operators, dendriform algebras and AYBEs

In this final section, we apply results from previous section to study the connection among the concepts that we have encountered so far: $\mathcal{O}$-operators, Rota-Baxter operators, dendriform algebra and AYBEs, as well as their relationship with relative differential operators that we will introduce next. We also obtain characterizations of dendriform algebras in terms of bimodules.

6.1. $\mathcal{O}$-operators and Rota-Baxter operators: the second connection

We show that an $\mathcal{O}$-operator, even though a generalization of Rota-Baxter operator, is equivalent to a Rota-Baxter operator on a larger algebra. Note the difference with the relationship in Section 2.3.

Proposition 6.1. *Let $(A, \cdot)$ be a **k**-algebra and let $(R, \circ, \ell, r)$ be an A-bimodule **k**-algebra. Let $\alpha : R \to A$ be a linear map. For a given $\lambda \in \mathbf{k}$, the following statements are equivalent.*

(a) The linear map α is an $\mathcal{O}$-operator of weight λ.

(b) The linear map

$$\hat{\alpha} : A \ltimes_{\ell,r} R \to A \ltimes_{\ell,r} R, \quad \hat{\alpha}(x, u) = (-\lambda x + \alpha(u), 0), \quad \forall x \in A, u \in R, \tag{47}$$

is a Rota-Baxter operator of weight λ.

(c) The linear map

$$-\lambda\mathrm{id} - \hat{\alpha} : A \ltimes_{\ell,r} R \to A \ltimes_{\ell,r} R, \quad (-\lambda\mathrm{id} - \hat{\alpha})(x, u) = (-\alpha(u), -\lambda u),$$
$$\forall x \in A, u \in R,$$

is a Rota-Baxter operator of weight λ.

Proof. (a) $\Leftrightarrow$ (b). Let $x, y \in A$ and $u, v \in R$. By Eq. (8) we have

$$\hat{\alpha}(x, u) * \hat{\alpha}(y, v) = (\alpha(u) \cdot \alpha(v) - \lambda \alpha(u) \cdot y - \lambda x \cdot \alpha(v) + \lambda^2 x \cdot y, 0),$$
$$\hat{\alpha}(\hat{\alpha}(x, u) * (y, v)) = (\lambda^2 x \cdot y - \lambda \alpha(u) \cdot y - \lambda \alpha(l(x)v) + \alpha(l(\alpha(u))v), 0),$$
$$\hat{\alpha}((x, u) * \hat{\alpha}(y, v)) = (\lambda^2 x \cdot y - \lambda x \cdot \alpha(v) - \lambda \alpha(ur(y)) + \alpha(ur(\alpha(v))), 0),$$
$$\hat{\alpha}((x, u) * (y, v)) = (-\lambda x \cdot y + (\alpha(l(x)v) + \alpha(ur(y)) + \alpha(u \circ v)), 0).$$

Therefore for any $x, y \in A$ and $u, v \in R$,

$$\alpha(u) \cdot \alpha(v) = \alpha(\ell(\alpha(u)v)) + \alpha(ur(\alpha(v))) + \lambda \alpha(u \circ v)$$

if and only if

$$\hat{\alpha}(x, u) * \hat{\alpha}(y, v) = \hat{\alpha}(\hat{\alpha}(x, u) * (y, v)) + \hat{\alpha}((x, u) * \hat{\alpha}(y, v)) + \lambda \hat{\alpha}((x, u) * (y, v)).$$

(b) $\Leftrightarrow$ (c) This follows from the well-known fact that $P : A \to A$ is a Rota-Baxter operator of weight λ on a **k**-algebra A if and only if $-\lambda \text{id} - P : A \to A$ is a Rota-Baxter operator of weight λ. $\quad\square$

Corollary 6.2. *Let $(A, \cdot)$ be a **k**-algebra and let $(R, \circ, \ell, r)$ be an A-bimodule **k**-algebra, both with finite **k**-dimensions. Let $\alpha : R \to A$ be an $\mathcal{O}$-operator of weight $\lambda \neq 0$ associated to $(R, \circ, \ell, r)$. Let*

$$R_\alpha(x, u) = \left(\frac{2}{\lambda}\hat{\alpha} + \text{id}\right)(x, u) = \left(\frac{2}{\lambda}\alpha(u) - x, u\right), \quad \forall x \in A, u \in R,$$

where $\hat{\alpha}$ is defined by Eq. (47). Then $(R_\alpha - R_\alpha^{21}) \pm (\text{id} + \text{id}^{21})$ is a solution of $AYBE$ in $(A \ltimes_{\ell,r} R) \ltimes_{R^*_{A \ltimes_{\ell,r} R}, L^*_{A \ltimes_{\ell,r} R}} (A \ltimes_{\ell,r} R)^*$ and $R_\alpha - R_\alpha^{21}$ is a skew-symmetric solution of $GAYBE$ in $(A \ltimes_{\ell,r} R) \ltimes_{R^*_{A \ltimes_{\ell,r} R}, L^*_{A \ltimes_{\ell,r} R}} (A \ltimes_{\ell,r} R)^*$.

Proof. By Proposition 6.1 and Corollary 2.11, $R_\alpha(x, u)$ is an extended $\mathcal{O}$-operator with modification id of mass -1. Then the first conclusion follows from Corollary 3.9.(d) and the second conclusion follows from Proposition 4.6.(d). $\quad\square$

A similar argument proves the following correspondence between $\mathcal{O}$-operators of weight zero and Rota-Baxter operators of any weight in a larger algebra.

Proposition 6.3. *Let A be a **k**-algebra and let (V, ℓ, r) be an A-bimodule. Let $\alpha : V \to A$ be a linear map. For given $\lambda, \mu \in \mathbf{k}$ with $\mu \neq 0$, the following statements are equivalent.*

(a) The linear map α is an $\mathcal{O}$-operator of weight zero.

(b) The linear map

$$\hat{\alpha}(x, u) = (\mu\alpha(u), -\lambda u), \quad \forall x \in A, u \in V, \tag{48}$$

is a Rota-Baxter operator of weight λ on $A \ltimes_{\ell,r} V$.

(c) The linear map

$$(-\lambda\mathrm{id} - \hat{\alpha})(x, u) = (-\lambda x - \mu\alpha(u), 0), \quad x \in A, u \in V,$$

is a Rota-Baxter operator of weight λ on $A \ltimes_{\ell,r} V$.

As a special case of Proposition 6.3 we obtain the following result from.[33] A linear map $\alpha : V \to A$ from an A-bimodule V to A is an $\mathcal{O}$-operator of weight zero (called a **generalized Rota-Baxter operator** in[33]) if and only if $\hat{\alpha}(x, u) := (\alpha(u), 0)$ is a Rota-Baxter operator of weight zero on $A \ltimes_{l,r} V$.

Corollary 6.4. *Let A be a $\mathbf{k}$-algebra and let (V, ℓ, r) be an A-bimodule, both with finite $\mathbf{k}$-dimensions. Let $\alpha : V \to A$ be an $\mathcal{O}$-operator of weight 0 associated to (V, ℓ, r).*

(a) Let $0 \neq \mu \in \mathbf{k}$ and let

$$\tilde{\alpha}(x, u) = (\mu\alpha(u), 0), \quad \forall x \in A, u \in V,$$

where $\mu \neq 0 \in \mathbf{k}$. Then $\tilde{\alpha} - \tilde{\alpha}^{21}$ is a skew-symmetric solution of AYBE in $(A \ltimes_{\ell,r} V) \ltimes_{R^*_{A \ltimes_{\ell,r}V}, L^*_{A \ltimes_{\ell,r}V}} (A \ltimes_{\ell,r} V)^*$.

(b) Let $\lambda, \mu \in \mathbf{k}$ with $\lambda \neq 0$ and let

$$R_\alpha(x, u) = \left(\frac{2}{\lambda}\hat{\alpha} + \mathrm{id}\right)(x, u) = \left(\frac{2\mu}{\lambda}\alpha(u) + x, -u\right), \quad \forall x \in A, u \in V,$$

where $\hat{\alpha}$ is defined by Eq. (48). Then $(R_\alpha - R_\alpha^{21}) \pm (\mathrm{id} + \mathrm{id}^{21})$ is a solution of AYBE in $(A \ltimes_{\ell,r} V) \ltimes_{R^*_{A \ltimes_{\ell,r}V}, L^*_{A \ltimes_{\ell,r}V}} (A \ltimes_{\ell,r} V)^*$ and $R_\alpha - R_\alpha^{21}$ is a skew-symmetric solution of GAYBE in $(A \ltimes_{\ell,r} V) \ltimes_{R^*_{A \ltimes_{\ell,r}V}, L^*_{A \ltimes_{\ell,r}V}} (A \ltimes_{\ell,r} V)^*$.

Proof. (a) By Proposition 6.3, $\tilde{\alpha}$ is a Rota-Baxter operator of weight zero. So the conclusion follows from Corollary 3.9.(b).

The proof of (b) is the same as the proof of Corollary 6.2. $\square$

When α is invertible, we further have

Proposition 6.5. *Let A be a $\mathbf{k}$-algebra and let (V, ℓ, r) be an A-bimodule. Let $\alpha : V \to A$ be an invertible linear map. Let $\lambda, \mu_1, \mu_2 \in \mathbf{k}$ with $\mu_1 \neq 0$ and $\mu_2 \neq \pm\lambda$. Then α is an $\mathcal{O}$-operator of weight zero if and only if*

$$\hat{\alpha}(x, u) = \left(\mu_1 \alpha(u) - \frac{\mu_2 + \lambda}{2} x, \frac{\lambda^2 - \mu_2^2}{4\mu_1} \alpha^{-1}(x) + \frac{\mu_2 - \lambda}{2} u \right), \ \forall x \in A, u \in V,$$

$$(49)$$

is a Rota-Baxter operator of weight λ on $A \ltimes_{\ell, r} V$.

Proof. Since α is invertible, we can assume that

$$\hat{\alpha}(x, u) = (ax + b\alpha(u), c\alpha^{-1}(x) + du), \ \forall x \in A, u \in V,$$

where $a, b, c, d \in \mathbf{k}$ are parameters. Let $x, y \in A$ and $u, v \in R$. Then by Eq. (8) we have

$$\hat{\alpha}(x, u) * \hat{\alpha}(y, v) = \big(b^2 \alpha(u) \cdot \alpha(v) + ab\alpha(u) \cdot y + abx \cdot \alpha(v) + a^2 x \cdot y,$$
$$l(ax + b\alpha(u))(c\alpha^{-1}(y) + dv) + (c\alpha^{-1}(x) + du)r(ay + b\alpha(v)) \big);$$

$$\hat{\alpha}(\hat{\alpha}(x, u) * (y, v)) = \big(a((ax + b\alpha(u)) \cdot y) + b\alpha(l(ax + b\alpha(u))v + (c\alpha^{-1}(x) + du)r(y)),$$
$$c\alpha^{-1}((ax + b\alpha(u)) \cdot y) + d(l(ax + b\alpha(u))v + (c\alpha^{-1}(x) + du)r(y)) \big);$$

$$\hat{\alpha}((x, u) * \hat{\alpha}(y, v)) = \big(a(x \cdot (ay + b\alpha(v))) + b\alpha(ur(ay + b\alpha(v)) + l(x)(c\alpha^{-1}(y) + dv)),$$
$$c\alpha^{-1}(x \cdot (ay + b\alpha(v))) + d(ur(ay + b\alpha(v)) + l(x)(c\alpha^{-1}(y) + dv)) \big);$$

$$\hat{\alpha}((x, u) * (y, v)) = \big(ax * y + b\alpha(l(x)v + ur(y)), c\alpha^{-1}(x * y) + d(l(x)v + ur(y)) \big).$$

Therefore when

$$a = -\frac{\mu_2 + \lambda}{2}, b = \mu_1, c = \frac{\lambda^2 - \mu_2^2}{4\mu_1}, d = \frac{\mu_2 - \lambda}{2},$$

we find that, for any $x, y \in A$ and $u, v \in V$,

$$\alpha(u) \cdot \alpha(v) = \alpha(\ell(\alpha(u)v)) + \alpha(ur(\alpha(v)))$$

if and only if

$$\hat{\alpha}(x, u) * \hat{\alpha}(y, v) = \hat{\alpha}(\hat{\alpha}(x, u) * (y, v)) + \hat{\alpha}((x, u) * \hat{\alpha}(y, v)) + \lambda \hat{\alpha}((x, u) * (y, v)).$$

$$\square$$

We then have the following direct consequence on AYBE by the same argument as for Corollary 6.4.

Corollary 6.6. *Let A be a $\mathbf{k}$-algebra and (V, ℓ, r) be an A-bimodule, both with finite $\mathbf{k}$-dimensions. Let $\alpha : V \to A$ be an invertible $\mathcal{O}$-operator of weight zero associated to (V, ℓ, r).*

(a) For $\mu_i \in \mathbf{k}$, $\mu_i \neq 0$, $i = 1, 2$, let

$$\tilde{\alpha}(x, u) = \left(\mu_1 \alpha(u) - \frac{\mu_2}{2} x, -\frac{\mu_2^2}{4\mu_1} \alpha^{-1}(x) + \frac{\mu_2}{2} u \right), \quad \forall x \in A, u \in V.$$

Then $\tilde{\alpha} - \tilde{\alpha}^{21}$ is a skew-symmetric solution of AYBE in $(A \ltimes_{\ell,r} V) \ltimes_{R^*_{A \ltimes_{\ell,r} V}, L^*_{A \ltimes_{\ell,r} V}} (A \ltimes_{\ell,r} V)^*$.

(b) For $\lambda, \mu_1, \mu_2 \in \mathbf{k}$ with $\lambda \neq 0$, $\mu_1 \neq 0$, $\mu_2 \neq \pm\lambda$ and for $\hat{\alpha}$ defined by Eq. (49), define

$$R_\alpha(x, u) = \left(\frac{2}{\lambda} \hat{\alpha} + \mathrm{id} \right)(x, u) = \left(\frac{2\mu_1}{\lambda} \alpha(u) - \frac{\mu_2}{\lambda} x, \frac{\lambda^2 - \mu_2^2}{2\lambda\mu_1} \alpha^{-1}(x) + \frac{\mu_2}{\lambda} u \right),$$

for all $x \in A, u \in V$. Then $(R_\alpha - R_\alpha^{21}) \pm (\mathrm{id} + \mathrm{id}^{21})$ is a solution of AYBE in $(A \ltimes_{\ell,r} V) \ltimes_{R^*_{A \ltimes_{\ell,r} V}, L^*_{A \ltimes_{\ell,r} V}} (A \ltimes_{\ell,r} V)^*$ and $R_\alpha - R_\alpha^{21}$ is a skew-symmetric solution of GAYBE in $(A \ltimes_{\ell,r} V) \ltimes_{R^*_{A \ltimes_{\ell,r} V}, L^*_{A \ltimes_{\ell,r} V}} (A \ltimes_{\ell,r} V)^*$.

6.2. Relative differential operators and Rota-Baxter operators

In this subsection, we introduce the concept of a relative differential operator and study its relationship with Rota-Baxter operators and AYBEs.

Definition 6.7. Let $f : A \to R$ be a linear map from a $\mathbf{k}$-algebra $(A, \cdot)$ to an A-bimodule $\mathbf{k}$-algebra $(R, \circ, \ell, r)$. If f satisfies

$$f(x \cdot y) = \ell(x)f(y) + f(x)r(y) + \lambda f(x) \circ f(y), \quad \forall x, y \in A,$$

then f is called a **relative differential operator of weight** λ.

Proposition 6.8. Let $f : A \to R$ be a linear map from a $\mathbf{k}$-algebra $(A, \cdot)$ to an A-bimodule $\mathbf{k}$-algebra $(R, \cdot, \ell, r)$. The following statements are equivalent.

(a) The linear map f is a relative differential operator of weight 1.
(b) The linear map

$$\hat{f} : A \ltimes_{\ell,r} R \to A \ltimes_{\ell,r} R, \quad \hat{f}(x, u) := (-x, -f(x)), \quad x \in A, u \in R, \quad (50)$$

is a Rota-Baxter operator of weight 1.
(c) The linear map

$$-\mathrm{id} - \hat{f} : A \ltimes_{\ell,r} R \to A \ltimes_{\ell,r} R, \quad (-\mathrm{id} - \hat{f})(x, u) := (0, f(x) - u),$$

$$x \in A, u \in R, \quad (51)$$

is a Rota-Baxter operator of weight 1.

Proof. (a) $\Leftrightarrow$ (b). Let $x, y \in A$ and $u, v \in R$. Then we have

$$\hat{f}(x, u) * \hat{f}(y, v) = (x * y, l(x)f(y) + f(x)r(y) + f(x) \circ f(y));$$
$$\hat{f}((x, u) * \hat{f}(y, v)) = (x * y, f(x * y)); \quad \hat{f}(\hat{f}(x, u) * (y, v)) = (x * y, f(x * y));$$
$$\hat{f}((x, u) * (y, v)) = (-x * y, -f(x * y)).$$

Therefore for any $x, y \in A$ and $u, v \in R$,

$$\hat{f}(x, u) * \hat{f}(y, v) = \hat{f}((x, u) * \hat{f}(y, v)) + \hat{f}(\hat{f}(x, u) * (y, v)) + \hat{f}((x, u) * (y, v))$$

if and only if

$$f(x * y) = l(x)f(y) + r(y)f(x) + f(x) \circ f(y).$$

(b) $\Leftrightarrow$ (c) This follows from the well-known fact that $P : A \to A$ is a Rota-Baxter operator of weight 1 if and only if $\tilde{P} := -\mathrm{id} - P : A \to A$ is a Rota-Baxter operator of weight 1. $\qquad\square$

Corollary 6.9. *Let $(A, \cdot)$ be a $\mathbf{k}$-algebra and let $(R, \circ, l, r)$ be an A-bimodule $\mathbf{k}$-algebra, both with finite $\mathbf{k}$-dimensions. Let $f : A \to R$ be a relative differential operator of weight 1. Define*

$$R_f(x, u) := (2\hat{f} + \mathrm{id})(x, u) = (-x, -2f(x) + u), \quad \forall x \in A, u \in R,$$

where $\hat{f}$ is defined by Eq. (50). Then $(R_f - R_f^{21}) \pm (\mathrm{id} + \mathrm{id}^{21})$ is a solution of AYBE in $(A \ltimes_{\ell, r} R) \ltimes_{R^*_{A \ltimes_{\ell, r} R}, L^*_{A \ltimes_{\ell, r} R}} (A \ltimes_{\ell, r} R)^*$ and $R_f - R_f^{21}$ is a skew-symmetric solution of GAYBE in $(A \ltimes_{\ell, r} R) \ltimes_{R^*_{A \ltimes_{\ell, r} R}, L^*_{A \ltimes_{\ell, r} R}} (A \ltimes_{\ell, r} R)^*$.

Proof. The proof is the same as the proof of Corollary 6.2. $\qquad\square$

When the A-bimodule $\mathbf{k}$-algebra is replaced by an A-bimodule, we have the following similar relation.

Proposition 6.10. *Let $f : A \to V$ be a linear map from a $\mathbf{k}$-algebra $(A, \cdot)$ to an A-bimodule (V, ℓ, r). Let $\lambda, \mu \in \mathbf{k}$ with $\lambda\mu \neq 0$. The following statements are equivalent.*

(a) The linear map f is a relative differential operator of weight 0.
(b) The linear map

$$\hat{f} : A \ltimes_{\ell, r} V \to A \ltimes_{\ell, r} V, \quad \hat{f}(x, u) = (-\lambda x, \mu f(x)), \quad x \in A, u \in V, \quad (52)$$

is a Rota-Baxter operator of weight λ.

(c) The linear map

$$-\lambda\mathrm{id} - \hat{f} : A \ltimes_{\ell,r} V \to A \ltimes_{\ell,r} V, \tag{53}$$
$$(-\lambda\mathrm{id} - \hat{f})(x, u) = (0, -\lambda u - \mu f(x)), \quad \forall x \in A, u \in V,$$

is a Rota-Baxter operator of weight λ.

Proof. It follows from a similar proof as the one for Proposition 6.8. $\quad\square$

Then similar to Corollary 6.9, we have

Corollary 6.11. *Let $(A, \cdot)$ be a k-algebra and (V, ℓ, r) be a bimodule, both with finite k-dimensions. Let $f : A \to V$ be a relative differential operator of weight 0. Let $\lambda, \mu \in k$ with $\lambda\mu \neq 0$ and let*

$$R_f(x, u) = \left(\frac{2}{\lambda}\hat{f} + \mathrm{id}\right)(x, u) = \left(-x, \frac{2\mu}{\lambda}f(x) + u\right), \quad \forall x \in A, u \in V, \tag{54}$$

where $\hat{f}$ is defined by Eq. (52). Then $(R_f - R_f^{21}) \pm (\mathrm{id} + \mathrm{id}^{21})$ is a solution of AYBE in $(A \ltimes_{\ell,r} V) \ltimes_{R^*_{A \ltimes_{\ell,r} V}, L^*_{A \ltimes_{\ell,r} V}} (A \ltimes_{\ell,r} V)^*$ and $R_f - R_f^{21}$ is a skew-symmetric solution of GAYBE in $(A \ltimes_{\ell,r} V) \ltimes_{R^*_{A \ltimes_{\ell,r} V}, L^*_{A \ltimes_{\ell,r} V}} (A \ltimes_{\ell,r} V)^*$.

6.3. *Characterizations of dendriform algebras in terms of bimodules and associativity*

We take the chance to present the following characterizations of dendriform algebras.

Proposition 6.12. *Let A be a k-vector space and let $\prec, \succ, \cdot : A \otimes A \to A$ be three binary operations. Set $\star = \prec + \succ + \cdot$.*

(a) $(A, \prec, \succ, \cdot)$ is a dendriform trialgebra if and only if $\star$ and $\cdot$ are associative products and $(A, \cdot, L_\succ, R_\prec)$ is an $(A, \star)$-bimodule k-algebra.

(b) Define

$$(x, a) \circ (y, b) = (x \star y, x \succ b + a \prec y + a \cdot b), \quad \forall x, y, a, b \in A. \tag{55}$$

Then $\circ$ is an associative product if and only if $(A, \prec, \succ, \cdot)$ is a dendriform trialgebra.

Proof. (a). "$\Longrightarrow$" For any $x, a, b \in A$, we have

$$L_\succ(x)(a \cdot b) = x \succ (a \cdot b) = (x \succ a) \cdot b = (L_\succ(x)a) \cdot b.$$

Thus, $L_\succ(x)(a \cdot b) = (L_\succ(x)a) \cdot b$. Other axioms defining an A-bimodule algebra can be verified in the similar way.

"$\Longleftarrow$" For any $x, a, b \in A$, we have that

$$x \succ (a \cdot b) = L_\succ(x)(a \cdot b) = (L_\succ(x)a) \cdot b = (x \succ a) \cdot b.$$

Thus, $x \succ (a \cdot b) = (x \succ a) \cdot b$. Other axioms defining a dendriform trialgebra can be verified in the similar way.

(b). By Proposition 2.3, Eq. (55) defines a **k**-algebra structure on the direct sum $A \oplus A$ of the underlying **k**-vector spaces of A and A (the semidirect sum) if and only if $\star$ is an associative product and $(A, \cdot, L_\succ, R_\prec)$ is an A-bimodule **k**-algebra. Hence the conclusion holds by Item (a). $\qquad\square$

Proposition 6.13. *Let A be a **k**-vector space and let $\star, \cdot : A \otimes A \to A$ be two associative products on A. Let $\ell, r : A \to \mathrm{End}_{\mathbf{k}}(A)$ be two linear maps such that $(A, \cdot, \ell, r)$ is an $(A, \star)$-bimodule **k**-algebra. Define two binary operations by*

$$x \succ y = \ell(x)y, \quad x \prec y = xr(y), \quad \forall x, y \in A. \tag{56}$$

(a) If

$$x \star y = \ell(x)y + xr(y) + x \cdot y = x \succ y + x \prec y + x \cdot y, \quad \forall x, y \in A, \tag{57}$$

then $(A, \succ, \prec, \cdot)$ is a dendriform trialgebra.

(b) Suppose that $\mathrm{Ann}_A A = 0$ and $(A, \succ, \prec, \cdot)$ is a dendriform trialgebra. Then Eq. (57) holds.

Proof. (a) In this case, id is an $\mathcal{O}$-operator of weight 1 of $(A, \star)$ associated to $(A, \cdot, \ell, r)$. Therefore the conclusion follows from Theorem 5.2.

(b) Since $(A, \cdot, \ell, r)$ is an $(A, \star)$-bimodule **k**-algebra and $(A, \succ, \prec, \cdot)$ is a dendriform trialgebra, for any $x, y, z \in A$, we have that

$$\ell(x \star y)z = \ell(x)(\ell(y)z) = x \succ (y \succ z) = (x \succ y + x \prec y + x \cdot y) \succ z$$
$$= \ell(x \succ y + x \prec y + x \cdot y)z.$$

Since $\mathrm{Ann}_A A = 0$, we have $x \star y = x \succ y + x \prec y + x \cdot y$. $\qquad\square$

We similarly obtain the following characterizations for dendriform dialgebras.

Proposition 6.14. *Let A be a **k**-vector space and let $\prec, \succ : A \otimes A \to A$ be two binary operations. Set $\star = \prec + \succ$.*

(a) The triple $(A, \prec, \succ)$ is a dendriform dialgebra if and only if $\star$ is an associative product and $(A, L_\succ, R_\prec)$ is an A-bimodule.

(b) Define

$$(x,a) \circ (y,b) = (x \star y, x \succ b + a \prec y), \quad \forall x,y,a,b \in A.$$

Then $\circ$ is an associative product if and only if $(A, \prec, \succ)$ is a dendriform dialgebra.

Proposition 6.15. *Let $(A, \star)$ be a $\mathbf{k}$-algebra and $\ell, r : A \to \mathrm{End}_{\mathbf{k}}(A)$ be two linear maps such that (A, ℓ, r) is an A-bimodule. Define two binary operations by*

$$x \succ y = \ell(x)y, \quad x \prec y = xr(y), \quad \forall x,y \in A.$$

(a) If

$$x \star y = \ell(x)y + xr(y) = x \succ y + x \prec y, \quad \forall x,y \in A, \qquad (58)$$

then $(A, \succ, \prec)$ is a dendriform dialgebra.

(b) Suppose that $\mathrm{Ann}_A A = 0$ and $(A, \succ, \prec)$ is a dendriform dialgebra. Then Eq. (58) holds.

6.4. Dendriform algebras and AYBEs

We now combine the connection between $\mathcal{O}$-operators and AYBEs and the connection between $\mathcal{O}$-operators and dendriform algebras to relate AYBEs to dendriform algebras. We first present a simple case. Let $(A, \prec, \succ)$ be a dendriform dialgebra and let $\star = \prec + \succ$. Note that id is an $\mathcal{O}$-operator of weight zero of $(A, \star)$ associated to the bimodule $(A, L_{\succ}, R_{\prec})$ (Proposition 6.14 (a)). Due to Corollary 3.9.(b), we have the following result.

Corollary 6.16. *Let $(A, \prec, \succ)$ be a dendriform dialgebra with finite $\mathbf{k}$-dimension. Then $\mathrm{id} - \mathrm{id}^{21}$ is a solution of AYBE in $A \ltimes_{R_{\prec}^*, L_{\succ}^*} A^*$, where $\mathrm{id} \in A \otimes A^*$ corresponds to the identity map $\mathrm{id} : A \to A$.*

For the dendriform dialgebra $(A, \succ, \prec)$, the linear map $\mathrm{id} : A \to A$ is also a relative differential operator of weight 0 from the $\mathbf{k}$-algebra $(A, \star)$ to the A-bimodule $(A, L_{\succ}, R_{\prec})$. Also note that id is invertible. Thus, from Proposition 6.3, Proposition 6.5 and Proposition 6.10 we obtain the following families of Rota-Baxter operators of weight λ on $A \ltimes_{L_{\succ}, R_{\prec}} A$:

$$\alpha(x,y) = (-\lambda x, \mu x), \quad \alpha(x,y) = (0, -\lambda y - \mu x), \quad \mu \in \mathbf{k}, \lambda\mu \neq 0, \qquad (59)$$

$$\alpha(x,y) = (-\lambda x - \mu y, 0), \quad \alpha(x,y) = (\mu y, -\lambda y), \quad 0 \neq \mu \in \mathbf{k}, \qquad (60)$$

$$\alpha(x,y) = \left(\mu_1 y - \frac{\mu_2 + \lambda}{2} x, \frac{\lambda^2 - \mu_2^2}{4\mu_1} x + \frac{\mu_2 - \lambda}{2} y \right), \quad \mu_1, \mu_2 \in \mathbf{k}, \mu_1 \neq 0, \mu_2 \neq \pm\lambda,$$

$$\qquad (61)$$

for all $x, y \in A$. In addition, when $\lambda = 0$ we get the following families of Rota-Baxter operators of weight 0 on $A \ltimes_{L_\succ, R_\prec} A$:

$$\alpha(x, y) = (-\mu y, 0), \quad \alpha(x, y) = (\mu y, 0), \quad 0 \neq \mu \in \mathbf{k}, \tag{62}$$

$$\alpha(x, y) = \left(\mu_1 y - \frac{\mu_2}{2} x, -\frac{\mu_2^2}{4\mu_1} x + \frac{\mu_2}{2} y \right), \quad \mu_1, \mu_2 \in \mathbf{k}, \mu_1 \neq 0, \mu_2 \neq \pm\lambda, \tag{63}$$

Thus from Proposition 3.9.(e), Proposition 4.6.(d) and Corollary 2.11 we obtain the following solutions of AYBE and GAYBE.

Corollary 6.17. *Let* $(A, \prec, \succ)$ *be a dendriform dialgebra with finite* $\mathbf{k}$-*dimension. Let* $\alpha : A \ltimes_{L_\succ, R_\prec} A \to A \ltimes_{L_\succ, R_\prec} A$ *be one of the operators defined in Eq. (59), Eq. (60) and Eq. (61), regarded as an element in* $\left((A \ltimes_{L_\succ, R_\prec} A) \ltimes_{R^*, L^*} (A \ltimes_{L_\succ, R_\prec} A)^* \right)^{\otimes 2}$. *Let* id *denote the identity operator on* $A \ltimes_{L_\succ, R_\prec} A$.

(a) *The elements* $\frac{2}{\lambda}(\alpha - \alpha^{21}) + 2\mathrm{id}$ *and* $\frac{2}{\lambda}(\alpha - \alpha^{21}) - 2\mathrm{id}^{21}$ *are solutions of AYBE in* $(A \ltimes_{L_\succ, R_\prec} A) \ltimes_{R^*, L^*} (A \ltimes_{L_\succ, R_\prec} A)^*$.

(b) *The element* $\frac{2}{\lambda}(\alpha - \alpha^{21}) + \mathrm{id} - \mathrm{id}^{21}$ *is a solution of GAYBE in* $(A \ltimes_{L_\succ, R_\prec} A) \ltimes_{R^*, L^*} (A \ltimes_{L_\succ, R_\prec} A)^*$.

Furthermore, by Proposition 3.9.(b) we obtain the following skew-symmetric solutions of AYBE.

Corollary 6.18. *Let* $(A, \prec, \succ)$ *be a dendriform dialgebra with finite* $\mathbf{k}$-*dimension. Let* $\alpha : A \ltimes_{L_\succ, R_\prec} A \to A \ltimes_{L_\succ, R_\prec} A$ *be one of the operators defined in Eq. (62) and Eq. (63), regarded as an element in* $\left((A \ltimes_{L_\succ, R_\prec} A) \ltimes_{R^*, L^*} (A \ltimes_{L_\succ, R_\prec} A)^* \right)^{\otimes 2}$. *Then* $\alpha - \alpha^{21}$ *is a skew-symmetric solution of AYBE in* $(A \ltimes_{L_\succ, R_\prec} A) \ltimes_{R^*, L^*} (A \ltimes_{L_\succ, R_\prec} A)^*$.

Let $(A, \prec, \succ, \cdot)$ be a dendriform trialgebra. From Proposition 6.12.(a) we know that $(A, \cdot, L_\succ, R_\prec)$ is an A-bimodule $\mathbf{k}$-algebra for $(A, \star)$, where $\star = \cdot + \prec + \succ$. Moreover, id $: A \to A$ is both an $\mathcal{O}$-operator (of weight 1) on A associated to the A-bimodule $\mathbf{k}$-algebra $(A, \cdot, L_\succ, R_\prec)$ and a relative differential operator (of weight 1) of A into the A-bimodule $\mathbf{k}$-algebra $(A, \cdot, L_\succ, R_\prec)$. Thus, from Proposition 6.1 and Proposition 6.8 we obtain the following families of Rota-Baxter operators of weight 1 on $A \ltimes_{L_\succ, R_\prec} A$:

$$\hat{\alpha}(x, y) = (-x + y, 0), \quad \hat{\alpha}(x, y) = -(y, y), \quad \hat{\alpha}(x, y) = -(x, x),$$
$$\hat{\alpha}(x, y) = (0, x - y), \tag{64}$$

for any $x, y \in A$. Thus from Proposition 3.9.(e), Proposition 4.6.(d) and Corollary 2.11, we obtain the following solutions of AYBE and GAYBE.

Corollary 6.19. *Let* $(A, \prec, \succ, \cdot)$ *be a dendriform trialgebra with finite* **k**-*dimension. Let* $\hat{\alpha} : A \ltimes_{L_\succ, R_\prec} A \to A \ltimes_{L_\succ, R_\prec} A$ *be one of the operators defined in Eq. (64), regarded as an element in* $\left((A \ltimes_{L_\succ, R_\prec} A) \ltimes_{R^*, L^*} (A \ltimes_{L_\succ, R_\prec} A)^* \right)^{\otimes 2}$. *Let* id *denote the identity operator on* $A \ltimes_{L_\succ, R_\prec} A$.

(a) *The elements* $2(\hat{\alpha} - \hat{\alpha}^{21}) + 2$id *and* $2(\hat{\alpha} - \hat{\alpha}^{21}) - 2$id^{21} *are solutions of* $AYBE$ *in* $(A \ltimes_{L_\succ, R_\prec} A) \ltimes_{R^*, L^*} (A \ltimes_{L_\succ, R_\prec} A)^*$.
(b) *The element* $2(\hat{\alpha} - \hat{\alpha}^{21}) +id-$id^{21} *is a solution of GAYBE in* $(A \ltimes_{L_\succ, R_\prec} A) \ltimes_{R^*, L^*} (A \ltimes_{L_\succ, R_\prec} A)^*$.

In fact, there are more Rota-Baxter operators that can be derived from dendriform algebras. For example,

$$\alpha(x, y) = -\lambda(x, 0) \quad \text{and} \quad \alpha(x, y) = -\lambda(0, y), \quad \forall x, y \in A,$$

are Rota-Baxter operators of weight λ on $A \ltimes_{L_\succ, R_\prec} A$ for a dendriform trialgebra $(A, \prec, \succ, \cdot)$.

Acknowledgements

C. Bai would like to thank the support by NSFC (10921061), NKBRPC (2006CB805905) and SRFDP (200800550015). L. Guo thanks the NSF grant DMS 0505445 for support and thanks the Chern Institute of Mathematics at Nankai University for hospitality.

References

1. M. Aguiar, Infinitesimal bialgebras, pre-Lie and dendriform algebras, in "Hopf Algebras", *Lecture Notes in Pure and Applied Mathematics* **237** (2004) 1-33.
2. M. Aguiar, Pre-Poisson algebras, *Lett. Math. Phys.* **54** (2000) 263-277.
3. M. Aguiar, On the associative analog of Lie bialgebras, *J. Algebra* **244** (2001) 492-532.
4. M. Aguiar and J.-L. Loday, Quadri-algebras, *J. Pure Appl. Algebra* **191** (2004) 205-221.
5. C. Bai, A unified algebraic approach to the classical Yang-Baxter equation, *J. Phys. A: Math. Theor.* **40** (2007) 11073-11082.
6. C. Bai, Double construction of Frobenius algebras, Connes cocycles and their duality, *J. Noncommutative Geometry* **4** (2010) 475-530.
7. C. Bai, $\mathcal{O}$-operators of Loday algebras and analogues of the classical Yang-Baxter equation, *Comm. Algebra* **38** (2010) 4277-4321.
8. C. Bai, L. Guo and X. Ni, Nonabelian generalized Lax pairs, the classical Yang-Baxter equation and PostLie algebras, *Comm. Math. Phys.* **297** (2010) 553-596.

9. C. Bai, L. Guo and X. Ni, $\mathcal{O}$-operators on associative algebras and associative Yang-Baxter equations, arXiv:0910.3261.

10. C. Bai, L. Guo and X. Ni, $\mathcal{O}$-operators on associative algebras and dendriform algebras, arXiv:1003.2432.

11. G. Baxter, An analytic problem whose solution follows from a simple algebraic identity, *Pacific J. Math.* **10** (1960) 731-742.

12. L. A. Bokut, Y. Chen and X. Deng, Gröbner-Shirshov bases for Rota-Baxter algebras, to appear in *J. Pure Appl. Algebra.*

13. M. Bordemann, Generalized Lax pairs, the modified classical Yang-Baxter equation, and affine geometry of Lie groups, *Comm. Math. Phys.* **135** (1990) 201-216.

14. P. Cartier, On the structure of free Baxter algebras, *Adv. Math.* **9** (1972) 253-265.

15. A. Connes, D. Kreimer, Renormalization in quantum field theory and the Riemann-Hilbert problem. I. The Hopf algebra structure of graphs and the main theorem, *Comm. Math. Phys.* **210** (2000) 249-273.

16. K. Ebrahimi-Fard, Loday-type algebras and the Rota-Baxter relation, *Lett. Math. Phys.* **61** (2002) 139-147.

17. K. Ebrahimi-Fard, On the associative Nijenhuis relation, *Elect. J. Comb.* **11** (2004), no. 1, Research Paper 38.

18. K. Ebrahimi-Fard and L. Guo, Rota-Baxter algebras and dendriform algebras, *J. Pure Appl. Algebra* **212** (2008) 320-339.

19. K. Ebrahimi-Fard, L. Guo and D. Kreimer, Integrable renormalization II: the general case, *Annales Henri Poincare* **6** (2005) 369-395.

20. L. Guo, WHAT IS a Rota-Baxter algebra, *Notice of Amer. Math. Soc.* **56** (2009) 1436-1437.

21. L. Guo, Operated semigroups, Motzkin paths and rooted trees, *J. Algebraic Combinatorics* **29** (2009) 35-62.

22. L. Guo and W. Keigher, Baxter algebras and shuffle products, *Adv. Math.* **150** (2000) 117-149.

23. L. Guo and W. Keigher, On differential Rota-Baxter algebras, *J. Pure and Appl. Algebra* **212** (2008) 522-540.

24. L. Guo and B. Zhang, Renormalization of multiple zeta values, *J. Algebra* **319** (2008) 3770-3809.

25. B.A. Kupershmidt, What a classical r-matrix really is, *J. Nonlinear Math. Phys.* **6** (1999) 448-488.

26. X. Li, D. Hou and C. Bai, Rota-Baxter operators on pre-Lie algebras, *J. Nonlinear Math. Phys.* **14** (2007) 269-289.

27. J.-L. Loday, Dialgebras, In: "Dialgebras and related operads", Lecture Notes in Math. **1763** (2002) 7-66.

28. J.-L. Loday, M. Ronco, Trialgebras and families of polytopes, in "Homotopy Theory: Relations with Algebraic Geometry, Group Cohomology, and Algebraic K-theory", *Contemp. Math.* **346** (2004) 369-398.

29. N. Reshetikhin and M. Semenov-Tian-Shansky, Quantum R-matrices and factorization problems, *J. Geom. Phys.* **5** (1998) 533-550.

30. G.-C. Rota, Baxter algebras and combinatorial identities I, II, *Bull. Amer.*

Math. Soc. **75** (1969) 325-329, 330-334.

31. G.-C. Rota, Baxter operators, an introduction, In: "Gian-Carlo Rota on Combinatorics", Joseph P.S. Kung, Editor, Birkhäuser, Boston, 1995, 504-512.

32. M.A. Semenov-Tyan-Shanskii, What is a classical R-matrix? *Funct. Anal. Appl.* **17** (1983) 259-272.

33. K. Uchino, Quantum analogy of Poisson geometry, related dendriform algebras and Rota-Baxter operators, *Lett. Math. Phys.* **85** (2008) 91-109.

34. V. N. Zhelyabin, Jordan bialgebras and their connection with Lie bialgebras, *Algebra and Logic* **36** (1997) 1-15.

IRREDUCIBLE WAKIMOTO-LIKE MODULES FOR THE AFFINE LIE ALGEBRA $\widehat{\mathfrak{gl}_n}$

YUN GAO*

*Department of Mathematics and Statistics, York University,
Toronto, Ontario M3J 1P3, Canada
E-mail: ygao@yorku.ca*

ZITING ZENG[†]

*School of Mathematical Science, Beijing Normal University,
Beijing, 100875 China
E-mail: zengzt@bnu.edu.cn*

In this paper, we used the free fields to construct a class of irreducible modules of the affine Lie algebra $\widehat{\mathfrak{gl}_n}$.

Keywords: Affine Lie algebra; free fields construction; irreducible modules.

1. Introduction

Affine Kac-Moody Lie algebras are (one-dimensional) universal central extensions of loop algebras which have been studied intensively by many researchers. People have constructed a great number of various irreducible modules for affine Lie algebras from different contexts. It is certainly not possible to classify all irreducible modules for an affine Kac-Moody Lie algebra. In some category of modules with certain good and natural properties, people are able to classify irreducible ones. For instance, Chari[1] classified all integrable irreducible modules for untwisted affine Kac-Moody Lie algebras with finite dimensional weight spaces. Chari-Pressley[3] classified the integrable irreducible modules for the twisted affine Kac-Moody Lie algebras. More precisely, suppose V is an integrable irreducible module for an affine

*The author gratefully acknowledges the grant support from the Natural Sciences and Engineering Research Council of Canada and the Chinese Academy of Science.
†The corresponding author. The author gratefully acknowledges the grant support from NNSF of China Nos. 10726014 and 10801010.

Kac-Moody Lie algebras, then V is a standard (so a highest weight) module if the central charge is positive; V is the dual (so a lowest weight) module of a standard module if the central charge is negative; V is a loop module if the central charge is zero. Thereafter Chari-Pressley[2] first constructed a continuous family of irreducible modules for affine Kac-Moody Lie algebras with infinite dimensional weight spaces. They achieved this by taking the tensor product of a standard module and an irreducible loop module.

In this paper, we will use Wakimoto's idea of free fields construction to obtain a family of irreducible modules for the affine Lie algebra $\widehat{\mathfrak{gl}_n}$ with infinite dimensional weight spaces. This approach is motivated by our previous works in Ref. 5 and 10 which traced back to Wakimoto's unpublished manuscript.[9] Note that the central charge $\neq 0$ in Chari-Pressley's construction[2] while the central charge $= 0$ in our construction. Thus our result can be seen as complementary to Chari-Pressley's result.

The free fields construction for affine Kac-Moody Lie algebras was introduced by Wakimoto[8] for the affine Lie algebra $\hat{\mathfrak{sl}}_2$ and in a great generality by Feigin-Frenkel in Ref.4 for the affine Lie algebras $\hat{\mathfrak{sl}}_n$.

The organization of this paper is as follows. In Section 2 we will first investigate the $\mathfrak{gl}_n$ and $\mathfrak{sl}_n$ case. The results in this section should be known to experts but not seen in any literature. Then we will deal with the affine Lie algebra $\widehat{\mathfrak{gl}_n}$ in Section 3.

2. Finite dimensional case

In this section we consider the general linear Lie algebra $\mathfrak{gl}_n$ and the special linear Lie algebra $\mathfrak{sl}_n$ over the complex field $\mathbb{C}$. As usual we denote $e_{i,j}$ to be the standard matrix units whose (i,j)-th entry is one and zero elsewhere.

Let $n \geq 2$ and
$$M = \mathbb{C}[x_2, x_3, \cdots, x_n]$$
be the polynomial ring with $n-1$ variables.

Define the following operators on M:

$$E_{i,1} = x_i, \ i = 2, \cdots, n;$$

$$E_{1,i} = \mu \frac{\partial}{\partial x_i} - x_2 \frac{\partial}{\partial x_2} \frac{\partial}{\partial x_i} - x_3 \frac{\partial}{\partial x_3} \frac{\partial}{\partial x_i} - \cdots - x_n \frac{\partial}{\partial x_n} \frac{\partial}{\partial x_i}, i = 2, \cdots, n;$$

$$E_{i,j} = x_i \frac{\partial}{\partial x_j}, \ i, j = 2, \cdots, n;$$

$$E_{1,1} = \mu - x_2 \frac{\partial}{\partial x_2} - x_3 \frac{\partial}{\partial x_3} - \cdots - x_n \frac{\partial}{\partial x_n};$$

Note that $E_{1,1} + E_{2,2} + \cdots + E_{n,n} = \mu$. The following result is known (easily verified).

Theorem 2.1. $\pi : \mathfrak{gl}_n(\mathbb{C}) \to \mathfrak{gl}(M)$ *with* $\pi(e_{i,j}) = E_{i,j}$, $i,j = 1, \cdots, n$ *is a representation for* $\mathfrak{gl}_n(\mathbb{C})$.

Write $L = \mathfrak{sl}_n(\mathbb{C})$. With this operation, M is a $\mathfrak{gl}_n(\mathbb{C})-$ module and thus a L module. Let H be the subalgebra of L which only contains those diagonal matrices, then H is the Cartan subalgebra of L.

Lemma 2.1. H *acts diagonally on* M *(so* M *is a weight module). If* $(i_2, i_3, \cdots, i_n) \neq (j_2, j_3, \cdots, j_n) \in \mathbb{Z}_+^{n-1}$, *where* $\mathbb{Z}_+$ *is the set of non-negative integers, then* $x_2^{i_2} x_3^{i_3} \cdots x_n^{i_n}$ *and* $x_2^{j_2} x_3^{j_3} \cdots x_n^{j_n}$ *belong to different weight spaces.*

Proof. Choose $\{e_{1,1} - e_{k,k}, k = 2, \cdots, n\}$ as a basis of H. Then

$$(E_{1,1} - E_{k,k}).x_2^{i_2} x_3^{i_3} \cdots x_n^{i_n}$$
$$= (\mu - i_2 - \cdots - 2i_k - i_{k+1} - \cdots - i_n)x_2^{i_2} x_3^{i_3} \cdots x_n^{i_n}.$$

So if $x_2^{i_2} x_3^{i_3} \cdots x_n^{i_n}$ and $x_2^{j_2} x_3^{j_3} \cdots x_n^{j_n}$ lie in the same weight space, then we have

$$\mu - 2i_2 - i_3 - \cdots - i_n = \mu - 2j_2 - j_3 - \cdots - j_n$$
$$\mu - i_2 - 2i_3 - \cdots - i_n = \mu - j_2 - 2j_3 - \cdots - j_n$$
$$\vdots$$
$$\mu - i_2 - i_3 - \cdots - 2i_n = \mu - j_2 - j_3 - \cdots - 2j_n.$$

That is

$$\begin{pmatrix} 2 & 1 & \cdots & 1 \\ 1 & 2 & \cdots & 1 \\ \vdots & \vdots & \cdots & \vdots \\ 1 & 1 & \cdots & 2 \end{pmatrix} \begin{pmatrix} i_2 - j_2 \\ i_3 - j_3 \\ \vdots \\ i_n - j_n \end{pmatrix} = 0.$$

Since

$$\begin{vmatrix} 2 & 1 & \cdots & 1 \\ 1 & 2 & \cdots & 1 \\ \vdots & \vdots & \cdots & \vdots \\ 1 & 1 & \cdots & 2 \end{vmatrix}_{(n-1)\times(n-1)} = n,$$

we have $i_2 = j_2, i_3 = j_3, \cdots, i_n = j_n$. $\qquad\square$

In what follows we set $h_i = e_{i,i} - e_{i+1,i+1}$ for $1 \leq i \leq n-1$ which form a basis for H. As usual we define linear functions $\epsilon_i(e_{j,j}) = \delta_{ij}$. Then the root system for $\mathfrak{sl}_n$ is

$$\Phi = \{\epsilon_i - \epsilon_j | 1 \leq i \neq j \leq n\}.$$

Let $\alpha_i = \epsilon_i - \epsilon_{i+1}, (1 \leq i \leq n-1)$ be a base for Φ. The fundamental dominant weight of L is

$$\epsilon_1, \epsilon_1 + \epsilon_2, \cdots, \epsilon_1 + \epsilon_2 + \cdots + \epsilon_{n-1}.$$

The Weyl group associated with Φ is the symmetric group S_n. So the weight of

$$x_2^{i_2} x_3^{i_3} \cdots x_n^{i_n}$$

is

$$(\mu - 2i_2 - i_3 \cdots - i_n, i_2 - i_3, i_3 - i_4, \cdots, i_{n-1} - i_n)$$
$$= \mu \cdot \mu_1 - (i_2 + \cdots + i_n)\alpha_1 - (i_3 + \cdots + i_n)\alpha_2 \cdots - i_n \alpha_{n-1}.$$

Theorem 2.2. *M is the highest weight module of L, with the highest weight λ such that*

$$(\lambda(h_1), \lambda(h_2), \cdots, \lambda(h_{n-1})) = (\mu, 0, \cdots, 0),$$

with the highest vector 1. Moreover, M is an irreducible module if and only if μ is not a nonnegative integer.

Proof. It is obvious that M is a highest weight module, with highest vector 1. Since $(e_{i,i} - e_{i+1,i+1}).1 = \mu$, if $i = 1$, and $(e_{i,i} - e_{i+1,i+1}).1 = 0$ otherwise.

If $U(\neq 0)$ is a submodule of M, we may assume that $x_2^{m_2} \cdots x_n^{m_n}$ is in U with $m = m_2 + \cdots + m_n \geq 1$ as every monomial lies in different weight space. Note that

$$\frac{1}{m_3! \cdots m_n!} E_{23}^{m_3} \cdots E_{2n}^{m_n} \cdot x_2^{m_2} \cdots x_n^{m_n} = x_2^{m_2 + \cdots m_n},$$

hence $x_2^{m_2 + \cdots m_n}$ is in U.

Since

$$E_{12} \cdot x_2^m = [m\mu - m(m-1)]x_2^{m-1} = m(\mu - (m-1))x_2^{m-1},$$

we have

$$E_{12}^m \cdot x_2^m = m!(\mu - (m-1)) \cdots (\mu - 1)\mu \cdot 1.$$

So if μ is not a nonnegative integer, then 1 is in U. Hence M is an irreducible module. $\square$

On the other hand, if μ is a nonnegative integer, then M has a nontrivial submodule generated by $x_2^{\mu+1}$ (containing all homogeneous polynomials with degree greater or equal to $\mu + 1$).

Theorem 2.3. *If μ is a nonnegative integer, the submodule $V(\mu)$ of M generated by $x_2^{\mu+1}$ is the maximal submodule of M. Moreover $V(\mu)$ itself is irreducible.*

Proof. If U is a submodule of M containing $V(\mu)$ as a submodule, then U must contain $x_2^{i_2} x_3^{i_3} \cdots x_n^{i_n}$ with $m := i_2 + i_3 \cdots + i_n \leq \mu$. Hence

$$x_2^{m_2 + \cdots m_n} = \frac{1}{m_3! \cdots m_n!} E_{23}^{m_3} \cdots E_{2n}^{m_n} \cdot x_2^{m_2} \cdots x_n^{m_n}$$

is in U, and

$$1 = \frac{1}{m!(\mu - (m-1)) \cdots (\mu - 1)\mu} E_{12}^m \cdot x_2^m$$

is in U. So $U = M$.

Also if $V(\mu)$ contains another highest weight vector different from $x_2^{\mu+1}$. It is obvious that the highest weight vector must have the form x_2^k, so if $k > \mu + 1$,

$$E_{1,2} \cdot x_2^k = k\mu - k(k-1)x_2^{k-1} = k(\mu + 1 - k)x_2^{k-1} \neq 0.$$

Hence x_2^k cannot be the highest weight vector if $k > \mu + 1$. So $V(\mu)$ is an irreducible module. $\qquad\square$

When μ is a nonnegative integer,
$V(\mu) = $ the $\mathbb{C}$-linear span of the monomials $x_2^{m_2}...x_n^{m_n}, m_2 + ... + m_n \geq \mu + 1$.

The quotient module $M/V(\mu)$ is a highest weight irreducible module of $\mathfrak{sl}_n$, with highest weight $(\mu, 0, \cdots, 0)$, hence it is finite dimensional. The basis for this quotient module is

$$\{x_2^{m_2}...x_n^{m_n} + V(\mu) | m_2 + ... + m_n \leq \mu\}.$$

The weight for $x_2^{m_2}...x_n^{m_n} + V(\mu)$ is

$$\mu \cdot \mu_1 + (m_2 + \cdots + m_n)\alpha_1 + \cdots + m_n\alpha_{n-1},$$

so the character formula for $M/V(\mu)$ is

$$Ch_{M/V(\mu)} = \sum_{\substack{m_i \geq 0 \\ \sum m_i \leq \mu}} e(\mu \cdot \mu_1)e((m_2 + \cdots + m_n)\alpha_1 + \cdots + m_n\alpha_{n-1})$$

In particular if $\mu = 0$, let

$$V = V(0)$$

be the subspace of M containing all the polynomials with degree ≥ 1. So

$$\{x_2^{i_2} x_3^{i_3} \cdots x_n^{i_n} \mid (i_2, \cdots, i_n) \in \mathbb{Z}_+^{n-1}, i_2 + \cdots + i_n \geq 1\}$$

form a basis of V.

Theorem 2.4. *(I) The character formula of M is*

$$ch\ (M) = \sum_{\alpha \in P} e(\mu \cdot \epsilon_1) e(\alpha),$$

where

$$P = \{\sum_{i=1}^{n-1} -m_i \alpha_i \mid m_1 \geq m_2 \geq \cdots \geq m_{n-1} \geq 0\}.$$

(II) If μ is a nonnegative integer, the character formula of $V(\mu)$ is

$$ch\ V(\mu) = \sum_{\alpha \in Q} e(\mu \cdot \epsilon_1) e(\alpha),$$

where

$$Q = \{\sum_{i=1}^{n-1} -m_i \alpha_i \mid m_1 \geq m_2 \geq \cdots \geq m_{n-1} \geq 0\ and\ m_1 \geq \mu + 1\}.$$

Proof. Since

$$E_{11}.x_2^{i_2} x_3^{i_3} \cdots x_n^{i_n} = (\mu - i_2 - i_3 - \cdots - i_n) x_2^{i_2} x_3^{i_3} \cdots x_n^{i_n},$$
$$E_{kk}.x_2^{i_2} x_3^{i_3} \cdots x_n^{i_n} = i_k x_2^{i_2} x_3^{i_3} \cdots x_n^{i_n}, k = 2, \cdots, n$$

we know that $x_2^{i_2} x_3^{i_3} \cdots x_n^{i_n}$ has the weight

$$(\mu - i_2 - i_3 - \cdots - i_n)\epsilon_1 + \sum_{k=2}^{n} i_k \epsilon_k$$
$$= \mu\epsilon_1 - (i_2 + i_3 + \cdots + i_n)(\epsilon_1 - \epsilon_2) - (i_3 + \cdots + i_n)(\epsilon_2 - \epsilon_3) - \cdots - i_n(\epsilon_{n-1} - \epsilon_n)$$
$$= \mu \cdot \epsilon_1 + (i_2 + i_3 + \cdots + i_n)(-\alpha_1) + (i_3 + \cdots + i_n)(-\alpha_2) + \cdots + i_n(-\alpha_{n-1})$$

the weight of $x_2^{i_2} x_3^{i_3} \cdots x_n^{i_n}$ lies in $\mu \cdot \mu_1 + P$.

On the other hand, for any $\lambda = \sum_{i=1}^{n-1} -m_i \alpha_i$ in P, we have

$$x_2^{m_1 - m_2} x_3^{m_2 - m_3} \cdots x_{n-1}^{m_{n-2} - m_{n-1}} x_n^{m_{n-1}}$$

whose weight is $\mu \cdot \mu_1 + \lambda$.

And from Lemma 2.1, we know that for any weight in P, the dimension of every weight space is 1. So we get the character formula of M and we prove (I).

For (II), it is similar to (I) if we notice that $x_2^{i_2} x_3^{i_3} \cdots x_n^{i_n}$ lies in $V(\mu)$ if and only if $i_2 + i_3 + \cdots + i_n \geq \mu + 1$. $\qquad\square$

Next let L_1 be the subalgebra of L generated by $< e_{i,j} \mid i \neq j, i, j = 2, \cdots, n >$. Then $L_1 = \mathfrak{sl}_{n-1}(\mathbb{C})$ whose columns and rows are indexed from 2 to n. Let $H_1 = H \cap L_1$ which is a Cartan subalgebra of L_1. Hence $\alpha_i = \epsilon_i - \epsilon_{i+1}, 2 \leq i \leq n - 1$ is a base for the root system Φ_1 of L_1.

Theorem 2.5. *Every homogeneous subspace M_m of M with degree m is an irreducible highest weight module of L_1, with the highest weight $m\epsilon_2$ and the highest weight vector x_2^m.*

Proof. Here $e_{ij}, i, j = 2, \cdots, n$ in L_1 acts on M_m. Indeed, for any $v \in M_m$,

$$E_{ij}.v = x_i \frac{\partial}{\partial x_j}.v$$

doesn't change the degree of v, M_m is a L_1-module.

If $m_2 + m_3 + \cdots + m_n = m$, $m_i \in \mathbb{Z}_+, i = 2, \cdots, n$,

$$x_2^{m_2} x_3^{m_3} \cdots x_n^{m_n} = \frac{1}{m_n! m_{n-1}! \cdots m_3!} (E_{32})^{m_3} (E_{42})^{m_4} \cdots (E_{n2}^{m_n}).x_2^m.$$

Therefore $M_m = U(L_1)x_2^m$.

If $i < j$, then $j > 2$, and

$$E_{ij}.x_2^m = x_i \frac{\partial}{\partial x_j}.x_2^m = 0.$$

For $2 \leq i < n$,

$$(E_{ii} - E_{i+1,i+1}).x_2^m = \begin{cases} m \cdot x_2^m, & i = 2 \\ 0, & i > 2 \end{cases}$$

Hence M_m is a highest weight module of L_1 with the highest weight vector x_2^m and the highest weight $m\epsilon_2$.

Since

$$E_{ii}.x_2^{m_2} x_3^{m_3} \cdots x_n^{m_n} = m_i x_2^{m_2} x_3^{m_3} \cdots x_n^{m_n},$$

for $i, j = 2, \cdots, n$, the weight of $x_2^{m_2} x_3^{m_3} \cdots x_n^{m_n}$ is $\sum_{i=2}^{n-1} m_i \epsilon_i$, so $x_2^{m_2} x_3^{m_3} \cdots x_n^{m_n}$ and $x_2^{i_2} x_3^{i_3} \cdots x_n^{i_n}$ lie in different weight space if $(m_2, m_3, \cdots, m_n) \neq (i_2, i_3, \cdots, i_n)$. And note that

$$x_2^{i_2 + i_3 + \cdots + i_n} = \frac{1}{i_3! \cdots i_n!} (E_{23})^{i_3} (E_{24})^{i_4} \cdots (E_{2n})^{i_n} x_2^{i_2} x_3^{i_3} \cdots x_n^{i_n},$$

M_m is an irreducible L_1-module. $\qquad\square$

Remark 2.1. Note that the $\mathfrak{sl}_n$-module

$$M = \oplus_{m=0}^{\infty} M_m$$

is obtained by taking a direct sum of finite dimensional irreducible $\mathfrak{sl}_{n-1}$-modules.

3. Affine case

In this section we consider the affine Lie algebra. Set

$$\widehat{\mathfrak{gl}_n(\mathbb{C})} = (\mathfrak{gl}_n(\mathbb{C}[t, t^{-1}])) \oplus \mathbb{C}c \oplus \mathbb{C}d$$

where c is the central element and d is the degree derivation. The Lie bracket is defined as

$$[A_1(t^{k_1}) + f_1 c + g_1 d, A_2(t^{k_2}) + f_2 c + g_2 d]$$
$$= [A_1, A_2](t^{k_1+k_2}) + g_1 k_2 A_2(t^{k_2}) - g_2 k_1 A_1(t^{k_1}) + tr(A_1 A_2)k_2 \delta_{k_1+k_2,0} c$$

for $A_1, A_2 \in \mathfrak{gl}_n$, $f_1, f_2, g_1, g_2 \in \mathbb{C}$, $k_1, k_2 \in \mathbb{Z}$.

Let

$$\widetilde{M} = \mathbb{C}[x_i(m), i = 1, \cdots, n, m \in \mathbb{Z}]$$

be a polynomial ring with infinitely many variables $x_i(m)$.

Define the following operators

$$E_{i,1}(m_1) = x_i(m_1), \ i = 2, \cdots, n;$$
$$E_{1,i}(m_1) = \mu \frac{\partial}{\partial x_i(-m_1)} - \sum_{m,m' \in \mathbb{Z}} x_2(m_1 + m + m') \frac{\partial}{\partial x_2(m)} \frac{\partial}{\partial x_i(m')}$$
$$- \sum_{m,m' \in \mathbb{Z}} x_3(m_1 + m + m') \frac{\partial}{\partial x_3(m)} \frac{\partial}{\partial x_i(m')} - \cdots$$
$$- \sum_{m,m' \in \mathbb{Z}} x_n(m_1 + m + m') \frac{\partial}{\partial x_n(m)} \frac{\partial}{\partial x_i(m')},$$
$$i = 2, \cdots, n;$$
$$E_{i,j}(m_1) = \sum_{m \in \mathbb{Z}} x_i(m_1 + m) \frac{\partial}{\partial x_j(m)}, \ i, j = 2, \cdots, n;$$
$$E_{1,1}(m_1) = \mu \delta_{m_1,0} - \sum_{m \in \mathbb{Z}} x_2(m_1 + m) \frac{\partial}{\partial x_2(m)}$$
$$- \sum_{m \in \mathbb{Z}} x_3(m_1 + m) \frac{\partial}{\partial x_3(m)} - \cdots$$
$$- \sum_{m \in \mathbb{Z}} x_n(m_1 + m) \frac{\partial}{\partial x_n(m)};$$
$$D = \sum_{i=1}^{n} \sum_{m \in \mathbb{Z}} m x_i(m) \frac{\partial}{\partial x_i(m)}.$$

Although most of the above operators are infinite sums, they are well defined as operators on $\widetilde{M}$.

Theorem 3.1. $\widehat{\pi} : \widehat{\mathfrak{gl}_n(\mathbb{C})} \to \mathfrak{gl}(\widetilde{M})$ *with*

$$\widehat{\pi}(e_{i,j}(t^m)) = E_{i,j}(m), i, j = 1, \cdots, n,$$
$$\widehat{\pi}(d) = D, \widehat{\pi}(c) = 0$$

is a representation for the affine Lie algebra $\widehat{\mathfrak{gl}_n(\mathbb{C})}$. In this way, $\widetilde{M}$ is the highest weight module of $\widehat{\mathfrak{gl}_n}$, where the Borel subalgebra consists of the upper triangular matrices (in $\mathfrak{gl}_n(\mathbb{C}[t, t^{-1}])$) together with d and c; the highest weight vector is 1 associated with the highest weight λ, where $\lambda(E_{1,1}(0)) = \mu$, and $\lambda(E_{i,i}(0)) = \lambda(d) = \lambda(c) = 0$, for $i = 2, \cdots, n$.

Proof. We check the commutation relations.

$$\begin{aligned}
[E_{i,1}(m_1), E_{1,i}(m_2)] &= -\mu\delta_{m_1,-m_2} + \sum x_2(m_2 + m + m_1)\frac{\partial}{\partial x_2(m)} \\
&\quad + \sum x_3(m_2 + m + m_1)\frac{\partial}{\partial x_3(m)} + \cdots \\
&\quad + \sum x_n(m_2 + m + m_1)\frac{\partial}{\partial x_n(m)} \\
&\quad + \sum x_i(m_2 + m_1 + m')\frac{\partial}{\partial x_i(m')} \\
&= E_{i,i}(m_1 + m_2) - E_{1,1}(m_1 + m_2)
\end{aligned}$$

for $2 \leq i \leq n$.

$$[E_{i,1}(m_1), E_{1,j}(m_2)] = \sum x_i(m_2 + m_1 + m')\frac{\partial}{\partial x_j(m')} = E_{i,j}(m_1 + m_2)$$

for $2 \leq i \neq j \leq n$.

$$[E_{i,1}(m_1), E_{k,l}(m_2)] = -\delta_{l,i}x_k(m_1 + m_2) = -\delta_{l,i}E_{k,1}(m_1 + m_2)$$

for $2 \leq i, k, l \leq n$.

$$[E_{i,1}(m_1), E_{1,1}(m_2)] = x_i(m_1 + m_2) = E_{i,1}(m_1 + m_2)$$

for $2 \leq i \leq n$.

$$\begin{aligned}
&[E_{1,i}(m_1), E_{1,i}(m_2)] \\
&= \left[-\sum x_i(m_1 + m + m')\frac{\partial}{\partial x_i(m)}\frac{\partial}{\partial x_i(m')}, -\sum x_i(m_2 + m + m')\frac{\partial}{\partial x_i(m)}\frac{\partial}{\partial x_i(m')}\right] \\
&= \sum x_i(m_1 + m + m_2 + m'' + m')\frac{\partial}{\partial x_i(m)}\frac{\partial}{\partial x_i(m'')}\frac{\partial}{\partial x_i(m')} \\
&\quad + \sum x_i(m_1 + m + m_2 + m'' + m')\frac{\partial}{\partial x_i(m)}\frac{\partial}{\partial x_i(m'')}\frac{\partial}{\partial x_i(m')} \\
&\quad - \sum x_i(m_1 + m + m_2 + m'' + m')\frac{\partial}{\partial x_i(m)}\frac{\partial}{\partial x_i(m'')}\frac{\partial}{\partial x_i(m')} \\
&\quad - \sum x_i(m_1 + m + m_2 + m'' + m')\frac{\partial}{\partial x_i(m)}\frac{\partial}{\partial x_i(m'')}\frac{\partial}{\partial x_i(m')} \\
&= 0
\end{aligned}$$

for $2 \leq i \leq n$.

If $i \neq j$, then

$$[E_{1,i}(m_1), E_{1,j}(m_2)]$$
$$= \left[-\sum x_j(m_1 + m + m') \frac{\partial}{\partial x_j(m)} \frac{\partial}{\partial x_i(m')} - \sum x_i(m_1 + m + m') \frac{\partial}{\partial x_i(m)} \frac{\partial}{\partial x_i(m')} , \right.$$
$$\left. -\sum x_i(m_2 + m + m') \frac{\partial}{\partial x_i(m)} \frac{\partial}{\partial x_j(m')} - \sum x_j(m_2 + m + m') \frac{\partial}{\partial x_j(m)} \frac{\partial}{\partial x_j(m')} \right]$$
$$= \sum x_j(m_1 + m_2 + m + m' + m'') \frac{\partial}{\partial x_j(m)} \frac{\partial}{\partial x_i(m'')} \frac{\partial}{\partial x_j(m')}$$
$$+ \sum x_j(m_1 + m_2 + m + m' + m'') \frac{\partial}{\partial x_i(m')} \frac{\partial}{\partial x_j(m)} \frac{\partial}{\partial x_j(m'')}$$
$$+ \sum x_i(m_1 + m_2 + m + m' + m'') \frac{\partial}{\partial x_i(m)} \frac{\partial}{\partial x_i(m'')} \frac{\partial}{\partial x_j(m')}$$
$$+ \sum x_i(m_1 + m_2 + m + m' + m'') \frac{\partial}{\partial x_i(m')} \frac{\partial}{\partial x_i(m)} \frac{\partial}{\partial x_j(m'')}$$
$$- \sum x_i(m_1 + m_2 + m + m' + m'') \frac{\partial}{\partial x_i(m)} \frac{\partial}{\partial x_j(m'')} \frac{\partial}{\partial x_i(m')}$$
$$- \sum x_i(m_1 + m_2 + m + m' + m'') \frac{\partial}{\partial x_j(m')} \frac{\partial}{\partial x_i(m)} \frac{\partial}{\partial x_i(m'')}$$
$$- \sum x_j(m_1 + m_2 + m + m' + m'') \frac{\partial}{\partial x_j(m)} \frac{\partial}{\partial x_j(m'')} \frac{\partial}{\partial x_i(m')}$$
$$- \sum x_j(m_1 + m_2 + m + m' + m'') \frac{\partial}{\partial x_j(m')} \frac{\partial}{\partial x_j(m)} \frac{\partial}{\partial x_i(m'')}$$
$$= 0$$

for $2 \leq i \neq j \leq n$.

$$[E_{1,i}(m_1), E_{i,l}(m_2)]$$
$$= \mu \frac{\partial}{\partial x_l(-m_1 - m_2)} - \sum x_2(m_1 + m_2 + m + m') \frac{\partial}{\partial x_2(m)} \frac{\partial}{\partial x_l(m')}$$
$$- \cdots$$
$$- \sum x_n(m_1 + m_2 + m + m') \frac{\partial}{\partial x_n(m)} \frac{\partial}{\partial x_l(m')}$$
$$- \sum x_i(m_1 + m_2 + m + m') \frac{\partial}{\partial x_i(m')} \frac{\partial}{\partial x_l(m)}$$
$$+ \sum x_i(m_1 + m_2 + m + m') \frac{\partial}{\partial x_l(m)} \frac{\partial}{\partial x_i(m')}$$
$$= E_{1,l}(m_1 + m_2, n_1 + n_2)$$

for $2 \leq i, l \leq n$.

If $i \neq k$, then

$$[E_{1,i}(m_1), E_{k,k}(m_2)]$$
$$= \left[-\sum x_k(m_1 + m + m') \frac{\partial}{\partial x_k(m)} \frac{\partial}{\partial x_i(m')}, \sum x_k(m_2 + m) \frac{\partial}{\partial x_k(m)} \right]$$
$$= -\sum x_k(m_1 + m_2 + m + m') \frac{\partial}{\partial x_i(m')} \frac{\partial}{\partial x_k(m)}$$
$$+ \sum x_k(m_1 + m_2 + m + m') \frac{\partial}{\partial x_k(m)} \frac{\partial}{\partial x_i(m')}$$
$$= 0$$

for $2 \leq i \neq k \leq n$.

If $i \neq k, l$, then

$$[E_{1,i}(m_1), E_{k,l}(m_2)]$$
$$= \left[-\sum x_k(m_1 + m + m') \frac{\partial}{\partial x_k(m)} \frac{\partial}{\partial x_i(m')} - \sum x_l(m_1 + m + m') \frac{\partial}{\partial x_l(m)} \frac{\partial}{\partial x_i(m')} , \right.$$
$$\left. \sum x_k(m_2 + m) \frac{\partial}{\partial x_l(m)} \right]$$
$$= -\sum x_k(m_1 + m_2 + m + m') \frac{\partial}{\partial x_i(m')} \frac{\partial}{\partial x_l(m)}$$
$$+ x_k(m_1 + m_2 + m + m') \frac{\partial}{\partial x_l(m)} \frac{\partial}{\partial x_i(m')}$$
$$= 0$$

for $2 \leq i, k, l \leq n$.

$$[E_{1,i}(m_1), E_{1,1}(m_2)]$$
$$= \mu \frac{\partial}{\partial x_i(-m_1-m_2)}$$
$$+ \sum_{k \neq i} \left[-\sum x_k(m_1 + m + m') \frac{\partial}{\partial x_k(m)} \frac{\partial}{\partial x_i(m')} , -\sum x_k(m_2 + m) \frac{\partial}{\partial x_k(m)} \right]$$
$$+ \sum_{k=2}^{n} \left[-\sum x_k(m_1 + m + m') \frac{\partial}{\partial x_k(m)} \frac{\partial}{\partial x_i(m')} , -\sum x_i(m_2 + m) \frac{\partial}{\partial x_i(m)} \right]$$
$$= \mu \frac{\partial}{\partial x_i(-m_1-m_2)} + \sum_{k \neq i} (\sum x_k(m_1 + m_2 + m + m') \frac{\partial}{\partial x_k(m)} \frac{\partial}{\partial x_i(m')}$$
$$- \sum x_k(m_1 + m_2 + m + m') \frac{\partial}{\partial x_k(m)} \frac{\partial}{\partial x_i(m')})$$
$$+ \sum_{k=2}^{n} \sum x_k(m_1 + m + m' + m_2) \frac{\partial}{\partial x_k(m)} \frac{\partial}{\partial x_i(m')}$$
$$- \sum x_k(m_1 + m + m' + m_2) \frac{\partial}{\partial x_k(m)} \frac{\partial}{\partial x_i(m')}$$
$$= \mu \frac{\partial}{\partial x_i(-m_1-m_2)} + \sum_{k=2}^{n} \sum x_k(m_1 + m_2 + m + m') \frac{\partial}{\partial x_k(m)} \frac{\partial}{\partial x_i(m')}$$
$$= E_{1,i}(m_1 + m_2)$$

for $2 \leq i \leq n$.

If $2 \leq i, j, k, l \leqq n$, then

$$[E_{i,j}(m_1, n_1), E_{k,l}(m_2, n_2)]$$
$$= \left[\sum x_i(m_1 + m) \frac{\partial}{\partial x_j(m)}, \sum x_k(m_2 + m) \frac{\partial}{\partial x_l(m)} \right]$$
$$= \delta_{jk} \sum x_i(m_1 + m_2 + m) \frac{\partial}{\partial x_l(m)} - \delta_{li} \sum x_k(m_1 + m_2 + m) \frac{\partial}{\partial x_j(m)}$$
$$= \delta_{jk} \sum x_i(m_1 + m_2 + m) \frac{\partial}{\partial x_l(m)} - \delta_{li} \sum x_k(m_1 + m_2 + m) \frac{\partial}{\partial x_j(m)}$$
$$= \delta_{jk} E_{i,l}(m_1 + m_2) - \delta_{li} E_{k,j}(m_1 + m_2).$$

If $2 \leq i \leq n$, then

$$[E_{i,i}(m_1, n_1), E_{1,1}(m_2, n_2)]$$
$$= \left[\sum x_i(m_1 + m) \frac{\partial}{\partial x_i(m)}, -\sum x_i(m_2 + m) \frac{\partial}{\partial x_i(m)} \right]$$
$$= -\sum x_i(m_1 + m_2 + m) \frac{\partial}{\partial x_i(m)} + \sum x_i(m_1 + m_2 + m) \frac{\partial}{\partial x_i(m)}$$
$$= 0.$$

If $2 \le i \ne j \le n$, then

$[E_{i,j}(m_1), E_{1,1}(m_2)]$

$= \left[\sum x_i(m_1 + m)\frac{\partial}{\partial x_j(m)}, -\sum x_i(m_2 + m)\frac{\partial}{\partial x_i(m)} - \sum x_j(m_2 + m)\frac{\partial}{\partial x_j(m)} \right]$

$= -\sum x_i(m_1 + m_2 + m)\frac{\partial}{\partial x_j(m)} + \sum x_i(m_1 + m_2 + m)\frac{\partial}{\partial x_j(m)}$

$= 0.$

Finally we have

$[E_{1,1}(m_1), E_{1,1}(m_2)]$

$\quad = \sum_{i=2}^n \left[-\sum x_i(m_1 + m)\frac{\partial}{\partial x_i(m)}, -\sum x_i(m_2 + m)\frac{\partial}{\partial x_i(m)} \right]$

$\quad = \sum_{i=2}^n (\sum x_i(m_1 + m_2 + m)\frac{\partial}{\partial x_i(m)} - \sum x_i(m_1 + m_2 + m)\frac{\partial}{\partial x_i(m)})$

$\quad = 0.$

The proof is completed. $\qquad\qquad\qquad\qquad\qquad\qquad\qquad\qquad\qquad\qquad$ $\square$

For $\widehat{\mathfrak{gl}_n}$, it is obvious that $\widetilde{M} = U(\widehat{\mathfrak{gl}_n}).1$, and

$E_{1,1}(0) \cdot x_2(m_{21}) \cdots x_2(m_{2k_2}) \cdots x_n(m_{n1}) \cdots x_n(m_{nk_n})$

$= (\mu - k_2 - k_3 \cdots - k_n)x_2(m_{21}) \cdots x_2(m_{2k_2}) \cdots x_n(m_{n1}) \cdots x_n(m_{nk_n}),$

$\quad E_{i,i}(0) \cdot x_2(m_{21}) \cdots x_2(m_{2k_2}) \cdots x_n(m_{n1}) \cdots x_n(m_{nk_n})$

$= k_i x_2(m_{21}) \cdots x_2(m_{2k_2}) \cdots x_n(m_{n1}) \cdots x_n(m_{nk_n})$

for $i = 2, \cdots, n$.

$D \cdot x_2(m_{21}) \cdots x_2(m_{2k_2}) \cdots x_n(m_{n1}) \cdots x_n(m_{nk_n})$

$= (\sum_{i=2}^n \sum_{j=1}^{k_i} m_{ij})x_2(m_{21}) \cdots x_2(m_{2k_2}) \cdots x_n(m_{n1}) \cdots x_n(m_{nk_n})$

In the following, the weight λ will be denoted as

$$(\lambda(e_{1,1}(1)), \lambda(e_{2,2}(1)), \cdots, \lambda(e_{n,n}(1)), \lambda(d)).$$

Theorem 3.2. *If $\mu \ne 0$, then $\widetilde{M}$ is an irreducible module.*

Proof. We first consider $\widehat{\mathfrak{gl}_2}$ case. Note that

$E_{1,2}(m^*).x_2(m_1) \cdots x_2(m_n)$

$= E_{1,2}(m^*)E_{2,1}(m_1) \cdots E_{2,1}(m_n).1$

$= (E_{1,1}(m^* + m_1) - E_{2,2}(m^* + m_1))E_{2,1}(m_2) \cdots E_{2,1}(m_n).1$

$\quad + E_{2,1}(m_1)E_{1,2}(m^*)E_{2,1}(m_2) \cdots E_{2,1}(m_n).1$

$= \cdots$

$= \sum_{i=1}^n E_{2,1}(m_1) \cdots \widehat{E_{2,1}(m_i)} \cdots E_{2,1}(m_n)(E_{1,1}(m^* + m_i) - E_{2,2}(m^* + m_i)).1$

$\quad - \sum_{i \ne j} E_{2,1}(m_1) \cdots \widehat{E_{2,1}(m_i)} \cdots \widehat{E_{2,1}(m_j)} \cdots E_{2,1}(m_n)E_{2,1}(m^* + m_i + m_j).1$

$= \mu \cdot \sum_{i=1}^n \delta_{m^*+m_i} E_{2,1}(m_1) \cdots \widehat{E_{2,1}(m_i)} \cdots E_{2,1}(m_n).1$

$\quad - \sum_{i \ne j} E_{2,1}(m_1) \cdots \widehat{E_{2,1}(m_i)} \cdots \widehat{E_{2,1}(m_j)} \cdots E_{2,1}(m_n)E_{2,1}(m^* + m_i + m_j).1$

where $\widehat{}$ means the term under is omitted.

Suppose U is a submodule of $\widetilde{M}$ generated by $v \neq 0$. If the weight of v is $(-1 + \mu, 1, m)$, then

$$v = ax_2(m), E_{1,2}(-m).v = \mu a.1,$$

so $U = \widetilde{M}$.

If the weight of v is $(-2 + \mu, 2, m)$, then

$$v = \sum a_i x_2(m_i) x_2(m - m_i)(\text{ finite sum and } a_i \neq 0),$$

with different m_i. We may assume that $m_i \leq \frac{m}{2}$, then

$$E_{1,2}(-m_j).v = (\mu a_j + \mu \delta_{m-2m_j,0} a_j - 2 \sum a_i) x_2(m - m_j),$$

and

$$E_{1,2}(k).v = (-2 \sum a_i) x_2(k + m)$$

for $k \neq -m_i$, and $n \neq -(m - m_i)$ for all i Hence

$$(\mu a_i + \mu \delta_{m-2m_j,0} a_j - 2 \sum a_i) x_2(m - m_1)$$

and

$$(-2 \sum a_i) x_2(n + m_1)$$

are both in U. Since at least one of them whose coefficient is nonzero, there must be one $x_2(k)$ which lies in U, from the discussion above, we see that $U = \widetilde{M}$.

If the weight of v is $(-n + \mu, n, m)$ with $n \geq 3$, then

$$v = \sum_{i=1}^{t} a_i x_2(m_{i_1}) \cdots x_2(m_{i_n}) = \sum_{i=1}^{t} A_i$$

(finite sum), with $m_{i_1} + \cdots m_{i_n} = m$, for all i.

Without loss of generality, we can assume

$$m_{i_1} \leq m_{i_2} \cdots \leq m_{i_n}, \text{ for all } i,$$

and

$$m_{1_{n-1}} + m_{1_n} \geq m_{2_{n-1}} + m_{2_n} \geq \cdots \geq m_{t_{n-1}} + m_{t_n}.$$

Consider $E_{12}(r).v$ with $r \neq -m_{i_j}$ for all i, j and big enough. Suppose for all $i \leq k$, $E_{1,2}(r).A_i$ can contribute one term in form $x_2(m_1) x_2(m_2) \cdots x_2(m_{n-1})$. That is

$$E_{1,2}(r).A_i = a_i(R - k_i x_2(m_1) x_2(m_2) \cdots x_2(m_{n-1}))$$

with

$$m_{n-1} = m_{1_{n-1}} + m_{1_n} + r,$$

here k_i's are positive integers (actually $k_i = 1$ for $i \geq 2$), R is the sum of the remaining terms. Hence the coefficient of $x_2(m_1)x_2(m_2) \cdots x_2(m_{n-1})$ is $-\sum a_i k_i$.

On the other hand, suppose m_{1_n} is the maximal among m_{i_n}, for $i \leq k$, consider $E_{12}(-m_{1_n}).v$, only for those $i \leq k$ can contribute the term

$$x_2(m_1)x_2(m_2) \cdots x_2(m_{n-2})x_2(m - m_{1n})$$

with coefficient $a_1 j \mu - \sum a_i k_i$, here $j > 0$.

So $E_{1,2}(-m_{1_n}).v$ and $E_{1,2}(r).v$ with $r \neq -m_{i_j}$ and big enough for all i, j, can not both be 0. The weight of $E_{1,2}(r).v$ is $(-n+1+\mu, n-1, m+r)$, by induction on n, we can get that $U = \widehat{M}$. We thus proved the case $\widehat{\mathfrak{gl}_2}$.

As for $\widehat{\mathfrak{gl}_n}$, since we have

$$E_{2,n}(a_n)^{k_n} \cdots E_{2,3}(a_3)^{k_3} x_2(m_{21}) \cdots x_2(m_{2k_2}) \cdots x_n(m_{n1}) \cdots x_n(m_{nk_n})$$
$$= (k_3! \cdots k_n!)x_2(m_{21}) \cdots x_2(m_{2k_2})x_2(m_{31} + a_3) \cdots x_2(m_{3k_3} + a_3)$$
$$\cdots x_2(m_{n1} + a_n) \cdots x_2(m_{nk_n} + a_n),$$

it reduces to the case of $\widehat{\mathfrak{gl}_2}$. The proof is therefore completed. $\qquad\square$

Remark 3.1. We use the degree derivation d in order to simply the proof. In fact, the theorem also holds true for loop algebra $\mathfrak{gl}_n(\mathbb{C}[t, t^{-1}])$.

We can see that $\mu = 0$, $\widetilde{M}$ is a reducible module, with the maximal submodule $\widetilde{V}$ containing all the polynomials with degree ≥ 1, and the quotient module $\widetilde{M}/\widetilde{V}$ is a trivial module.

Theorem 3.3. *If $\mu = 0$, there is a $(\mathfrak{gl}_n \otimes \mathbb{C}[t, t^{-1}])$-module homomorphism*

$$\varphi : \widetilde{V} \to V(0) \otimes \mathbb{C}[t, t^{-1}]$$

defined as

$$\varphi(x_2(m_{21}) \cdots x_2(m_{2k_2}) \cdots x_n(m_{n1}) \cdots x_n(m_{nk_n})) = x_2^{k_2} \cdots x_n^{k_n} \otimes t^{\sum_{i,k_j} m_{ik_j}}$$

Proof. We need to check that

$$\varphi(E_{i,j}(m).v) = E_{i,j} \otimes t^m.\varphi(v)$$

for any $0 < i, j < n$, $m \in \mathbb{Z}$, and

$$v = x_2(m_{21}) \cdots x_2(m_{2k_2}) \cdots x_n(m_{n1}) \cdots x_n(m_{nk_n}).$$

For $i \neq 1$, we have

$$E_{i,1}(m).x_2(m_{21}) \cdots x_2(m_{2k_2}) \cdots x_n(m_{n1}) \cdots x_n(m_{nk_n})$$
$$= x_i(m)x_2(m_{21}) \cdots x_2(m_{2k_2}) \cdots x_n(m_{n1}) \cdots x_n(m_{nk_n}).$$

It follows that

$$\varphi(E_{i,1}(m).v) = x_2^{k_2} \cdots x_i^{k_i+1} \cdots x_n^{k_n} \otimes t^{\sum_{i,k_j} m_{ik_j}+m} = E_{i,1} \otimes t^m.\varphi(v).$$

Since

$$E_{1,i}(m).x_2(m_{21}) \cdots x_2(m_{2k_2}) \cdots x_n(m_{n1}) \cdots x_n(m_{nk_n})$$
$$= -\sum_{r=2}^{n}\sum_{l=1}^{k_i}\sum_{j=1}^{k_r} x_2(m_{21}) \cdots \widehat{x_r(m_{rj})} \cdots \widehat{x_i(m_{il})}$$
$$\cdots x_n(m_{nk_n})x_r(m_{rj}+m+m_{il}),$$

we have

$$\varphi(E_{1,i}(m).v)$$
$$= -(\sum_{r=1,r\neq i}^{n} k_i k_r x_2^{k_2} \cdots x_i^{k_i-1} \cdots x_n^{k_n} + k_i(k_i-1)x_2^{k_2} \cdots x_i^{k_i-1} \cdots x_n^{k_n}$$
$$\otimes t^{\sum_{i,k_j} m_{ik_j}+m})$$
$$= E_{1,i} \otimes t^m.\varphi(v).$$

For $i, j \neq 1$,

$$E_{i,j}(m).x_2(m_{21}) \cdots x_2(m_{2k_2}) \cdots x_n(m_{n1}) \cdots x_n(m_{nk_n})$$
$$= \sum_{l=1}^{k_j} x_2(m_{21}) \cdots \widehat{x_j(m_{jl})} \cdots x_n(m_{nk_n})x_i(m+m_{jl})$$

and

$$\varphi(E_{i,j}(m).v) = k_j x_2^{k_2} \cdots x_i^{k_i+1} \cdots x_j^{k_j-1} \cdots x_n^{k_n} \otimes t^{\sum_{i,k_j} m_{ik_j}+m}$$
$$= E_{i,j} \otimes t^m.\varphi(v).$$

Since $E_{1,1}(m) = [E_{1,2}(m), E_{2,1}(0)] + E_{2,2}(m)$, we can conclude that

$$\varphi(E_{1,1}(m).v) = E_{1,1} \otimes t^m.\varphi(v),$$

so we are done. $\qquad\square$

It is clear that every different weight space of $\widetilde{V}$ is mapped to different one-dimensional space in $V(0) \otimes \mathbb{C}[t, t^{-1}]$. Hence we have

Corollary 3.1. $\widetilde{V}$ has a maximal submodule V', generated by all

$$x_2(0)^{k_2} x_3(0)^{k_3} \cdots x_i(m)x_i(0)^{k_i-1} \cdots x_n(0)^{k_n}$$
$$- x_2(m_{21}) \cdots x_2(m_{2k_2}) \cdots x_n(m_{n1}) \cdots x_n(m_{nk_n}) \quad (1)$$

for $k_j \geq 0$, $j = 2, \cdots, n$, and i is the minimal such that $k_i \geq 1$, and $\sum_{i,k_j} m_{ik_j} = m$.

Proof. From the theorem above, for $k_j \geq 0$, $j = 2, \cdots, n$, and i is the minimal such that $k_i \geq 1$, and $\sum_{i,k_j} m_{ik_j} = m$,

$$x_2(0)^{k_2} x_3(0)^{k_3} \cdots x_i(m) x_i(0)^{k_i-1} \cdots x_n(0)^{k_n}$$
$$- x_2(m_{21}) \cdots x_2(m_{2k_2}) \cdots x_n(m_{n1}) \cdots x_n(m_{nk_n})$$

lies in the kernel of φ.

On the other hand, if v is in the kernel of φ, then $v = v_1 + \cdots + v_m$ with v_i in different weight space. Hence every v_i is in kernel of φ.

Suppose $v_i = \sum b_j B_j$ with every B_j has the form

$$x_2(m_{21}) \cdots x_2(m_{2k_2}) \cdots x_n(m_{n1}) \cdots x_n(m_{nk_n}),$$

hence $\sum b_j = 0$. So

$$v_j = \sum b_j (x_2(0)^{k_2} x_3(0)^{k_3} \cdots x_i(m) x_i(0)^{k_i-1} \cdots x_n(0)^{k_n} - B_j)$$

is a linear combination of (1).

So the kernel of φ is V'. Since $V(0) \otimes \mathbb{C}[t, t^{-1}]$ is irreducible module, so V' is the maximal submodule of $\widetilde{V}$. $\qquad \square$

References

1. V. Chari, *Integrable representations of affine Lie-algebras, Invent. Math.* **85** (1986), pp. 317–335.

2. V. Chari and A. Pressley, *A new family of irreducible, integrable modules for affine Lie algebras, Math. Ann.* **277** (1987), pp. 543–562.

3. V. Chari and A. Pressley, *Integrable representations of twisted affine Lie algebras, J. Algebra* **113** (1988), pp. 438–464.

4. B. Feigin and E. Frenkel, *Affine Kac-Moody algebras and semi-infinite flag manifolds, Comm. Math. Phys.* **128** (1990), pp. 161–189.

5. Y. Gao and Z. Zeng, *Hermitian representations of the extended affine Lie algebra $\widehat{\mathfrak{gl}_2(\mathbb{C}_q)}$, Adv. Math.* **207** (2006), pp. 244–265.

6. V. Kac, *Infinite dimensional Lie algebras*, 3rd edition (Cambridge University Press, 1990).

7. R. V. Moody and A. Pianzola, *Lie algebras with triangular decomposition* (John Wiley, 1995, New York).

8. M. Wakimoto, *Fock representations of the affine Lie algebra $A_1^{(1)}$, Comm. Math. Phys.* **104** (1986), pp. 605–609.

9. M. Wakimoto, *Extended affine Lie algebras and a certain series of Hermitian representations, Preprint* (1985).

10. Z. Zeng, *Unitary representations of the extended affine Lie algebra $\widetilde{\mathfrak{gl}_3(\mathbb{C}_q)}$, Pacific J. Math.* **233** (2007), pp. 481–509.

VERMA MODULES
OVER GENERIC EXP-POLYNOMIAL LIE ALGEBRAS

XIANGQIAN GUO* and XUEWEN LIU[†]

*Department of Mathematics, Zhengzhou University,
Zhengzhou, Henan, 450001, P. R. China*
E-mail: guoxq@zzu.edu.cn
[†] *E-mail: liuxw@zzu.edu.cn*

KAIMING ZHAO

*Department of Mathematics, Wilfrid Laurier University,
Waterloo, ON, Canada N2L 3C5
and
College of Mathematics and Information Science,
Hebei Normal (Teachers) University,
Shijiazhuang, Hebei, 050016 P. R. China
Email: kzhao@wlu.ca*

In this paper, we first study properties of generic exp-polynomial functions on $\mathbb{Z}^n$. Then we use these properties to deduce a criterion for Verma modules to be irreducible over generic exp-polynomial Lie algebras relative to a total order.

Keywords: exp-polynomial function; exp-polynomial Lie algebra; Verma module.

1. Introduction

In recent years, representation theory for higher rank infinite dimensional Lie algebras such as extended affine Lie algebras, generalized Virasoro algebras, toroidal Lie algebras and quantum torus Lie algebras has attracted extensive attentions from many mathematicians (See Ref. 1–19). This is because representation theory of infinite dimensional Lie algebras has many important applications in mathematics and physics.

These Lie algebras, graded by $\mathbb{Z}^n (n > 1)$, do not possess the classical triangular decomposition as defined in Ref. 20. But the standard construction of the highest weight modules still gives a lot of interesting irreducible modules. Actually many authors constructed Verma modules over some of these algebras with respect to some total orders on $\mathbb{Z}^n$, and irreducibilities of the Verma modules were investigated (See Ref. 10,15,19,21, etc.).

Most of the above-mentioned Lie algebras are exp-polynomial Lie algebras or central extensions of exp-polynomial Lie algebras (Def. 2.2) which were introduced in Ref. 22, where the authors proved that an irreducible ($\mathbb{Z}$-graded) highest weight module with an exp-polynomial highest weight over an exp-polynomial Lie algebra has all weight spaces finite dimensional, which generalized a result for polynomial algebras in Ref. 3. The modules constructed in Ref. 22 have been used in several places, including classification of irreducible Harish-Chandra modules Ref. 12, and helped understand some affine Kac-Moody algebra modules (See Ref. 23).

Various examples of exp-polynomial Lie algebras were displayed in Ref. 22. The class of exp-polynomial Lie algebras is too huge for us to study their representations uniformly. We will make some restrictions so that the problem will be solvable.

In this paper, we introduce the concept of generic exp-polynomial functions and generic exp-polynomial Lie algebras of depth 1 (Def. 2.1–2.3). Such Lie algebras include many useful algebras, for example, higher rank Virasoro algebras, generalized Block algebras, loop-Virasoro algebras and generalized Virasoro-like algebras (See Examples 1-4). We give a criterion for irreducibility of Verma modules over generic exp-polynomial Lie algebras of depth 1 relative to some total orders (Theorem 2.1). Our main result (Theorem 2.1) and applications to these examples are explained in Section 2, yielding new results for several classes of Lie algebras.

As we will see, the reducibility of Verma modules depends on the exp-polynomial functions (defining the algebra and module) and the total order. So we deduce some properties of generic exp-polynomial functions in Section 3, which are critical to the proof of our result on Verma modules. In Section 4, we first collect some known results on total orders on $G = \mathbb{Z}^n$. Then we give the proof of our main results. The main idea and techniques are from Ref. 10 and Ref. 24, but our arguments in the present paper are more conceptional and less computational, thanks to properties of generic exp-polynomial functions established in Section 3.

Throughout this paper, we denote by $\mathbb{C}$, $\mathbb{R}$ and $\mathbb{Q}$ the fields of complex numbers, real numbers and rational numbers respectively. Denote by $\mathbb{Z}$, $\mathbb{Z}_+$ and $\mathbb{N}$ the set of integers, non-negative integers and positive integers respectively. All Lie algebras in this paper are over $\mathbb{C}$. For any $\alpha \in \mathbb{C}^n$, we denote by $\alpha(j)$ the j-th entry of α, that is $\alpha = (\alpha(1), \alpha(2), ..., \alpha(n))$. For any $\alpha \in \mathbb{Z}^n, \beta \in \mathbb{C}^n$, we write $\beta^\alpha = \beta(1)^{\alpha(1)}\beta(2)^{\alpha(2)}...\beta(n)^{\alpha(n)}$ for short, whenever the equation makes sense.

2. Main results and applications

In this section, we first recall the definition of exp-polynomial functions and exp-polynomial Lie algebras from Ref. 22. Then we introduce the concept of generic exp-polynomial function and generic exp-polynomial Lie algebras and state our main results with some examples. We will also apply our results to some concrete Lie algebras.

Definition 2.1. The **algebra of exp-polynomial functions** in $\alpha \in \mathbb{Z}^n$ is the commutative associative algebra of functions $f(\alpha) : \mathbb{Z}^n \to \mathbb{C}$ generated as an algebra by functions $\alpha(j)$ and $a^{\alpha(j)}$ for various $a \in \mathbb{C}^* = \mathbb{C} \setminus \{0\}$, $j = 1, ..., n$.

An exp-polynomial function may be written as a finite sum

$$f(\alpha) = \sum_{(\beta,\gamma) \in B \times \Gamma} a_{\beta\gamma} \alpha^\gamma \beta^\alpha, \tag{1}$$

where Γ and B are finite subsets of $\mathbb{Z}_+^n$ and $(\mathbb{C}^*)^n$ respectively. Here we have set $0^0 = 1$. We call such f **nonzero as an exp-polynomial function in** α, denoted by $f \neq 0$, if there is some $a_{\beta\gamma} \neq 0$. The **support** of f is defined as $\operatorname{supp}(f) = \{(\beta, \gamma) \mid a_{\beta\gamma} \neq 0\}$. Let P_f be the multiplicative subgroup of $\mathbb{C}^*$ generated by all $\beta(i)$ for all $(\beta, \gamma) \in \operatorname{supp}(f)$.

An exp-polynomial function f on $\mathbb{Z}^n$ is called **generic** if P_f is a free subgroup of $\mathbb{C}^*$.

Let us recall the lexicographical order on $\mathbb{Z}^n$: Given any $\beta, \beta' \in \mathbb{Z}^n$ with $\beta = (\beta(1), ..., \beta(n))$ and $\beta' = (\beta'(1), ..., \beta'(n))$, $\beta \succ \beta'$ if and only if there is $0 \leq s \leq n$, such that $\beta(s) > \beta'(s)$ and $\beta(t) = \beta'(t)$, $\forall\ s \leq t \leq n$.

From now on we assume that f is a generic exp-polynomial function on $\mathbb{Z}^n$ as in (1). We fix a total order $\prec$ on P_f which is compatible with the multiplication. There are actually many of such total orders on P_f. Later we will take the one we need.

Now we can use the above two total orders to define a total order on $P_f \times \mathbb{Z}_+^n$: $(\beta, \alpha) \prec (\beta', \alpha')$ if $\alpha \prec \alpha'$ or $\alpha = \alpha'$ and $\beta \prec \beta'$. In the next we always use this total order on $P_f \times \mathbb{Z}_+^n$.

Suppose that (γ_1, β_1) is the maximal elements in $\operatorname{supp}(f)$. We call (γ_1, β_1) the **degree** of f and $a_{\gamma_1, \beta_1} \beta_1^\alpha \alpha^{\gamma_1}$ the **highest term** of f, denoted by $\deg(f)$ and $\operatorname{ht}(f)$ respectively.

If f_1, f_2 are two nonzero exp-polynomial function on $\mathbb{Z}^n$ with $P_{f_1} P_{f_2}$ being a free multiplicative group, then we can fix a total order on $P_{f_1} P_{f_2}$.

The highest term of $f_1 f_2$ is just the product of the highest terms of f_1 and f_2. So if f_1 and f_2 are both nonzero, then $f_1 f_2$ is also nonzero.

Definition 2.2. Let $\mathcal{L} = \bigoplus_{\alpha \in \mathbb{Z}^n} \mathcal{L}_\alpha$ be a $\mathbb{Z}^n$ graded Lie algebra and T be an index set. Then $\mathcal{L}$ is said to be an **exp-polynomial Lie algebra** if $\mathcal{L}$ has a homogeneous spanning set $\{L_{t,\alpha} \in \mathcal{L}_\alpha | t \in T, \alpha \in \mathbb{Z}^n\}$ and there exists a family of exp-polynomial functions $\{f_{tr}^s(\alpha, \beta) | t, r, s \in T\}$ in (α, β) and where for each t, r the set $\{s | f_{tr}^s(\alpha, \beta) \neq 0\}$ is finite, such that for all nonzero $L_{t,\alpha}, L_{r,\beta}$ (we do not care about 0 elements in defining the brackets),

$$[L_{t,\alpha}, L_{r,\beta}] = \sum_{s \in T} f_{tr}^s(\alpha, \beta) L_{s,\alpha+\beta}, \qquad \forall\, t, r \in T \; \alpha, \beta \in \mathbb{Z}^n. \qquad (2)$$

Moreover, we call $\mathcal{L}$ a **generic exp-polynomial Lie algebra**, if all $f_{tr}^s(\alpha, \beta)$ occurred in (2) are generic exp-polynomial functions.

We remark that this definition is a little more general than the original one in Ref. 22.

Definition 2.3. Suppose $|T| = 1$ in Def. 2.2 and let $\mathcal{L}_\alpha = \mathbb{C} L_\alpha$, $\forall\, \alpha \in \mathbb{Z}^n$ with only finitely many $L_\alpha = 0$. Let $\widehat{\mathcal{L}} = \mathcal{L} \oplus \sum_{i=1}^m \mathbb{C} C_i$ be an m-dimensional central extension of the generic exp-polynomial Lie algebra $\mathcal{L}$, such that

$$[L_\alpha, L_\beta] = f(\alpha, \beta) L_{\alpha+\beta} + \delta_{\alpha+\beta,0} \sum_{i=1}^m f_i(\alpha) C_i, \qquad \forall\, \alpha, \beta \in \mathbb{Z}^n, \qquad (3)$$

where each f_i is a generic exp-function on $\mathbb{Z}^n$. We can also extend $\widehat{\mathcal{L}}$ by adding some linearly independent derivations $D_j, j = 1, ..., l$ to get $\widetilde{\mathcal{L}} = \widehat{\mathcal{L}} \oplus \sum_{j=1}^l \mathbb{C} D_j$, such that $[D_j, L_\alpha] = g_j(\alpha) L_\alpha$, and $[D_j, C_i] = 0$, where g_j are linearly independent additive functions. It is easy to see that $\mathcal{L}$ and $\widehat{\mathcal{L}}$ are generic exp-polynomial Lie algebras if we choose suitable spanning sets. We call them **generic exp-polynomial Lie algebras of depth 1**. In general, $\widetilde{\mathcal{L}}$ is not an exp-polynomial Lie algebra.

Example 2.1. Let $G \cong \mathbb{Z}^n$ be an additive subgroup of $\mathbb{C}$. The **rank n Virasoro algebra** $\mathbf{Vir}[G]$ is the Lie algebra with a $\mathbb{C}$ basis $\{C, d_g | g \in G\}$ and Lie bracket:

$$[d_g, d_h] = (h - g) d_{g+h} + \delta_{g+h,0} \frac{g^3 - g}{12} C, \quad [C, d_g] = 0, \quad \forall\, g, h \in G.$$

It is clear that $\mathbf{Vir}[G]$ is a generic exp-polynomial Lie algebra of depth 1.

Example 2.2. Let G be an additive subgroup of $\mathbb{C}$, and $\phi, \psi : G \to \mathbb{C}$ be two $\mathbb{Z}$ linearly independent additive maps with $\ker(\phi) \cap \ker(\psi) = 0$. Albert and Frank Ref. 25 defined the Lie algebra $\mathcal{L}(G, \phi, \psi)$ with a basis $\{L_\alpha | \alpha \in G\}$ and Lie bracket

$$[L_\alpha, L_\beta] = (\phi(\alpha - \beta) + (\phi \wedge \psi)(\alpha, \beta)) L_{\alpha+\beta},$$

where $(\phi \wedge \psi)(\alpha, \beta) = \phi(\alpha)\psi(\beta) - \phi_\beta \psi(\alpha)$. Let Z be the center of $\mathcal{L}$. Then $Z \subset [\mathcal{L}, \mathcal{L}]$, $\dim(Z) \leq 1$ and $\dim(L/[L, L]) \leq 1$. It was proved by Ref. 26 that the Lie algebra $[\mathcal{L}, \mathcal{L}]/Z$ is simple, which they called **generalized Block algebra**. When $Z = 0$ and $[\mathcal{L}, \mathcal{L}] = \mathcal{L}$, the algebra is called **Block algebra** Ref. 27. It is easy to see that the generalized Block algebras are generic exp-polynomial Lie algebras of depth 1.

Example 2.3. The **loop-Virasoro algebra** $\widetilde{\mathcal{L}}$ is the Lie algebra that is the tensor product of the Virasoro Lie algebra Vir and the Laurent polynomial algebra $\mathbb{C}[t^{\pm 1}]$, i.e., $\widetilde{\mathcal{L}} = \text{Vir} \otimes \mathbb{C}[t^{\pm 1}]$ with a basis $\{C \otimes t^j, d_i \otimes t^j | i, j \in \mathbb{Z}\}$ subject to the commutator relations:

$$[d_i \otimes t^j, d_k \otimes t^l] = (k - i)\left(d_{i+k} \otimes t^{j+l}\right) + \delta_{i+k,0} \frac{i^3 - i}{12}\left(C \otimes t^{j+l}\right),$$

$$[d_i \otimes t^j, C \otimes t^l] = 0.$$

It is clear that $\mathcal{L} = \widetilde{\mathcal{L}}/Z_1$ is a generic exp-polynomial Lie algebra of depth 1, where $Z_1 = \bigoplus_{i \in \mathbb{Z}, i \neq 0} C \otimes t^i$.

Example 2.4. Let $G_i \cong \mathbb{Z}^{n_i}, i = 1, 2$ be two additive subgroup of $\mathbb{C}$. The **rank (n_1, n_2) Virasoro-like algebra** $\mathcal{L}[G_1, G_2]$ is the Lie algebra with a $\mathbb{C}$ basis $\{C_1, C_2, L_{x_1, x_2} | (x_1, x_2) \in G_1 \times G_2\}$ and subject to the following commutator:

$$[L_{x_1, x_2}, L_{y_1, y_2}] = (x_2 y_1 - x_1 y_2) L_{x_1 + y_1, x_2 + y_2} + \delta_{x_1 + y_1, 0} \delta_{x_2 + y_2, 0}(x_1 C_1 + x_2 C_2),$$

where $x_i, y_i \in G_i, i = 1, 2$. Take $G = G_1 \oplus G_2$, then $\mathcal{L}[G_1, G_2]$ is a generic exp-polynomial Lie algebra of depth 1.

Now we recall some concepts on total orders of abelian groups. Suppose that G is an abelian group and that $\preceq$ is a total order on G which is compatible with the addition, i.e., $\alpha \preceq \beta$ implies $\alpha + \gamma \preceq \beta + \gamma$, for all $\alpha, \beta, \gamma \in G$. We have obvious meanings for the symbols $\prec, \succ, \succeq$. In this paper, **all total orders on an additive group G are total orders compatible with the addition of G**. We usually denote totally ordered set by the pair $(G, \prec)$.

The **rank** of G, denoted by rank(G), is defined as the maximal $r \in I\!N$ such that there exist $g_1, ..., g_r \in G$ which are linearly independent over $\mathbb{Z}$.

Let $(G, \prec)$ be a totally ordered abelian group and H is a subgroup of G, then we call $(H, \prec)$ a totally ordered subgroup of $(G, \prec)$, where the order on H is inherited from G. $(G, \prec)$ is called **archimedean** if for any $0 \prec g_1 \prec g_2 \in G$ there is some $n \in I\!N$ such that $g_2 \prec ng_1$. $(G, \prec)$ is called **discrete** if there exists a minimal element $a_0 \succ 0$ in G. $(G, \prec)$ is called **dense** if for any $0 \prec g_1$, there exist g_2 such that $0 \prec g_2 \prec g_1$, and is called **completely dense** if $(G, \prec)$ is dense and any rank 2 total ordered subgroup of $(G, \prec)$ is dense. It is easy to see that a completely dense group is an archimedean group. The only archimedean group that is not completely dense is $\mathbb{Z}$.

For any totally ordered abelian groups $(G, \prec)$ and $(H, \prec)$, we have the product order on the product group $G \times H$ as follows: $(g_1, h_1,) \prec (g_2, h_2)$ if $h_1 \prec h_2$ or $h_1 = h_2$ and $g_1 \prec g_2$. The resulting totally ordered group is denoted as $(G \times H, \prec)$. A totally ordered abelian group is called **indecomposable with respect to the order** $\prec$ if it is not isomorphic to a product of any two nonzero total ordered abelian groups.

Now we introduce some special classes of total orders on the abelian group $\mathbb{Z}^n$. Choose $0 < \theta_i \in I\!R, i = 1, 2, ..., n$, which are linearly independent over $\mathbb{Q}$, and we make the identification $\mathbb{Z}^n = \bigoplus_{i=1}^n \mathbb{Z}\theta_i$. Then we have the totally ordered abelian group $(\mathbb{Z}^n, \prec) = (\bigoplus_{i=1}^n \mathbb{Z}\theta_i, <)$ with the natural order inherited from $I\!R$. We refer these orders as **standard total orders** on $\mathbb{Z}^n$, and denote these total ordered abelian groups by $\mathbb{Z}(\theta_1, ..., \theta_r)$. Note also that $\mathbb{Z}(\theta_1, ..., \theta_r)$ is archimedean. It is obvious that total orders on $\mathbb{Z}$ are either the natural order or the reverse order, which are not dense.

Let $\mathcal{L}, \widehat{\mathcal{L}}$ and $\widetilde{\mathcal{L}}$ be as defined in Definition 2.3. Suppose that $\prec$ is a total order on $G = \mathbb{Z}^n$.

Let $G_+ = \{\alpha \in G | \alpha \succ 0\}$, $G_- = \{\alpha \in G | \alpha \prec 0\}$. Denote $\mathcal{L}_+ = \widehat{\mathcal{L}}_+ = \widetilde{\mathcal{L}}_+ = \bigoplus_{\alpha \succ 0} \mathcal{L}_\alpha = \bigoplus_{\alpha \succ 0} \mathbb{C} L_\alpha$, $\mathcal{L}_- = \widehat{\mathcal{L}}_- = \widetilde{\mathcal{L}}_- = \bigoplus_{\alpha \prec 0} \mathcal{L}_\alpha = \bigoplus_{\alpha \prec 0} \mathbb{C} L_\alpha$ and $\mathcal{L}_0 = \mathbb{C} L_0$, $\widehat{\mathcal{L}}_0 = \mathbb{C} L_0 \oplus \sum_{i=1}^m \mathbb{C} C_i$, $\widetilde{\mathcal{L}}_0 = \mathbb{C} L_0 \oplus \sum_{i=1}^m \mathbb{C} C_i \oplus \sum_{j=1}^l \mathbb{C} D_j$. For convenience, we also denote $\mathcal{L}(S) = \bigoplus_{\alpha \in S} \mathcal{L}_\alpha$ for any $S \subset G$, and similar for $\widehat{\mathcal{L}}(S)$ and $\widetilde{\mathcal{L}}(S)$.

Now we can define Verma modules over $\mathcal{L}, \widehat{\mathcal{L}}$ and $\widetilde{\mathcal{L}}$.

For linear map $\varphi : \mathcal{L}_0 \to \mathbb{C}$, we can define a 1-dimensional $\mathcal{L}_+ \oplus \mathcal{L}_0$ module $V = \mathbb{C} v_0$ by $\mathcal{L}_+ v_0 = 0, L_0 v_0 = \varphi(L_0)v_0$. Then we have the induced $\mathcal{L}$-module $M(\prec, \varphi) = U(\mathcal{L}) \otimes_{\mathcal{L}_+ \oplus \mathcal{L}_0} V$, called **the Verma module with respect to** $(\prec, \varphi)$. The vector v_0 is called a **highest weight vector** of $M(\prec, \varphi)$. We also write $\varphi(L_0) = h$ and $M(\prec, \varphi) = M(\prec, h)$ for convention.

Similarly, for linear map $\widehat{\varphi} : \widehat{\mathcal{L}}_0 \to \mathbb{C}$ and $\widetilde{\varphi} : \widetilde{\mathcal{L}}_0 \to \mathbb{C}$, we can define the Verma modules $M(\prec, \widehat{\varphi}) = M(\prec, h, c_1, .., c_m)$ and $M(\prec, \widetilde{\varphi}) = M(\prec, h, c_1, .., c_m, d_1, .., d_l)$, over $\widehat{\mathcal{L}}$ and $\widetilde{\mathcal{L}}$ respectively, where $c_i = \widehat{\varphi}(C_i), i = 1, .., m$ and $d_j = \widetilde{\varphi}(D_j), j = 1, ..., l$.

Clearly, $M(\prec, h, c_1, .., c_m, d_1, .., d_l)$ is irreducible over $\widetilde{\mathcal{L}}$ if and only if $M(\prec, h, c_1, .., c_m)$ is irreducible over $\widehat{\mathcal{L}}$, and $M(\prec, h)$ is irreducible over $\mathcal{L}$ if and only if $M(\prec, h, 0, ..., 0)$ is irreducible over $\widehat{\mathcal{L}}$. Thus we may only consider the Verma module $M(\prec, h, c_1, ..., c_m)$ over $\widehat{\mathcal{L}}$. We will denote $c = (c_1, ..., c_m)$ for short.

Let $Z(\widehat{\mathcal{L}})$ be the center of $\widehat{\mathcal{L}}$. If $L_\alpha \in Z(\widehat{\mathcal{L}})$ for some $\alpha \in G_-$, then $L_\beta L_\alpha v_0 = L_\alpha L_\beta v_0 = 0$ for any $\beta \succ 0$. Then the irreducibility of $M(\prec, h, c)$ implies $L_\alpha v_0 = 0$, for otherwise $L_\alpha v_0$ would be another highest weight vector of $M(\prec, h, c)$. Thus we may only consider the case $Z(\widehat{\mathcal{L}}) \subset \widehat{\mathcal{L}}_0$.

We have the natural G-gradation $M(\prec, h, c) = M_0 \bigoplus_{\lambda \in G_-} M_\lambda$ with $v_0 \in M_0$, where

$$M_\lambda = \sum_{\alpha_1 + \alpha_2 + \cdots + \alpha_s = \lambda} \mathbb{C} \, L_{\alpha_1} L_{\alpha_2} \cdots L_{\alpha_s} v_0.$$

When $M_\lambda \neq 0$, λ is called a **weight** of $M(\prec, h, c)$, M_λ is called a weight subspace and nonzero vectors in M_λ are called weight vectors.

Now we can present our main results:

Theorem 2.1. Let $\widehat{\mathcal{L}}$ be a generic exp-polynomial Lie algebra of depth 1 as defined in Def. 2.3. Suppose that $[\widehat{\mathcal{L}}, \widehat{\mathcal{L}}] = \widehat{\mathcal{L}}$ and $Z(\widehat{\mathcal{L}}) \subset \widehat{\mathcal{L}}_0$.

(1) Suppose that $\prec$ is completely dense. Then $M(\prec, h, c)$ is irreducible if and only if $f_{h,c}(\alpha) = f(\alpha, -\alpha)h + \sum_{i \in I} c_i f_i(\alpha)$ is nonzero as a function in α. If $f_{h,c}(\alpha) = 0$ then $M' = \bigoplus_{\alpha \in G_-} M_\alpha$ is an irreducible submodule of codimension one.

(2) Suppose that $(G, \prec) \cong (H, \prec) \times (K, \prec)$, with $H \prec K$ and $(H, \prec)$ being standard and nonzero. Then $M(\prec, h, c)$ is irreducible if and only if the $\widehat{\mathcal{L}}(H)$-Verma module $M_H(\prec, h, c) = U(\widehat{\mathcal{L}}(H)) \otimes_{\widehat{\mathcal{L}}(H)_+ \oplus \widehat{\mathcal{L}}(H)_0} \mathbb{C} \, v_0$ is irreducible.

By Theorem 2.1 , we deduce the following corollary:

Corollary 2.1. Let $\widehat{\mathcal{L}}$ be a generic exp-polynomial Lie algebra of depth 1 with $[\widehat{\mathcal{L}}, \widehat{\mathcal{L}}] = \widehat{\mathcal{L}}$ and $Z(\widehat{\mathcal{L}}) \subset \widehat{\mathcal{L}}_0$, and $M(\prec, h, c)$ be a Verma module defined as above. Suppose that $(G, \prec) \cong (H, \prec) \times (K, \prec)$, with $(H, \prec)$ standard and nonzero (note that K can be 0).

(1) *Suppose $\prec$ is dense. Then $M(\prec, h, c)$ is irreducible if and only if $f_{h,c} = f(\alpha, -\alpha)h + \sum_{i \in I} c_i f_i(\alpha)$ is nonzero as a function in $\alpha \in H$. If $f_{h,c} = 0$ as an function in $\alpha \in H$, then $M' = \bigoplus_{\alpha \in G_-} M_\alpha$ is an irreducible submodule of codimension one.*

(2) *Suppose $\prec$ is discrete. Then $M(\prec, h, c)$ is irreducible if and only if $M_a(\prec, h, c)$ is irreducible over $\widehat{\mathcal{L}}(\mathbb{Z}a)$, where $H = \mathbb{Z}a$ and a is the smallest positive element of G.*

Proof. If $\prec$ is discrete, then $H \simeq \mathbb{Z}$. The result (2) follows directly from Theorem 2.1 (2). If $\prec$ is dense, then H has rank at least 2. Consequently H is completely dense, and (1) follows from Theorem 2.1 (1) and (2). $\square$

Remark 2.1. The proof of Theorem 2.1 is also valid for G being an arbitrary subgroup of $\mathbb{C}$, if we suppose that $f(\alpha, \beta)$ is a polynomial function in (α, β). So the results in Theorem 2.1 and Corollary 2.1 can be generalized to any arbitrary G being an additive subgroup of $\mathbb{C}$, provided that the generic exp-polynomial function $f(\alpha, \beta)$ is indeed a polynomial function.

Example 2.5. Let $\mathrm{Vir}[G]$ be the rank n Virasoro algebra in Example 1. Then Corollary 2.1 gives Theorem 3.1 in Ref. 10 for $G \cong \mathbb{Z}^n$. Moreover, from Remark 2.1, the general result of Theorem 3.1 in Ref. 10 can be deduced from our proofs.

Example 2.6. Let $\mathcal{L}[G]$ be the rank n Block algebra in Example 2. Theorem 2.1 and Corollary 2.1 gives the criterion for irreducibility of Verma modules over these Block Lie algebras relative to some total order $\prec$ on $G \cong \mathbb{Z}^n$. Also the result can be generalized to the case for G being any arbitrary subgroup of $\mathbb{C}$, due to Remark 2.1.

Example 2.7. Let $\mathcal{L}(\mathrm{Vir})$ be the loop-Virasoro algebra in Example 3 and $\mathcal{L} = \mathcal{L}(\mathrm{Vir})/Z_1$. We can similarly define $\mathbb{Z} \times \mathbb{Z}$ Verma modules $M_{\mathcal{L}(\mathrm{Vir})}(\prec, h, c)$ and $M_{\mathcal{L}}(\prec, h, c)$ over $\mathcal{L}(\mathrm{Vir})$ and $\mathcal{L}$ respectively, relative to some total order $\prec$ on $\mathbb{Z}^2$.

Suppose that v is the highest weight vector of $M_{\mathcal{L}(\mathrm{Vir})}(\prec, h, c)$. It is obvious that $(C \otimes t^j)v$ is a highest weight vector other than v for any $0 \neq j \in \mathbb{Z}$. The irreducibility of $M_{\mathcal{L}(\mathrm{Vir})}(\prec, h, c)$ implies that $(C \otimes t^j)v = 0$, and hence $(C \otimes t^j)M_{\mathcal{L}(\mathrm{Vir})}(\prec, h, c)$ for any $j \neq 0$, that is, $M_{\mathcal{L}(\mathrm{Vir})}(\prec, h, c)$ can be viewed as an $\mathcal{L}$ module, which is just $M_{\mathcal{L}}(\prec, h, c)$.

Thus we have that $M_{\mathcal{L}(\mathrm{Vir})}(\prec, h, c)$ is irreducible over $\mathcal{L}(\mathrm{Vir})$ if and only if $(C \otimes t^j)M_{\mathcal{L}(\mathrm{Vir})}(\prec, h, c) = 0, j \neq 0$ and $M_{\mathcal{L}}(\prec, h, c)$ is irreducible over $\mathcal{L}$.

Since $\mathcal{L}$ is a generic exp-polynomial Lie algebra of depth 1, we can get the criterion for irreducibility of Verma modules over $\mathcal{L}$ and $\mathcal{L}(\text{Vir})$ relative to some total order $\prec$ on $\mathbb{Z}^2$, by Theorem 2.1.

Example 2.8. Let $\mathcal{L}[G_1, G_2]$ be the Virasoro-like algebra of rank (n_1, n_2) in Example 4. Then Theorem 2.1 and Corollary 2.1 give the criterion for irreducibility of Verma modules over $\mathcal{L}[G_1, G_2]$. When $G_1 = G_2 = \mathbb{Z}$, our results give Theorem 3.1(1) in Ref. 21.

3. Properties on generic exp-polynomial functions

In this section, we will give some properties on generic exp-polynomials, which are crucial to the proof of the main results.

For any $a \in \mathbb{R}$, we denote by $\bar{a}$ the representative of $a + \mathbb{Z}$ in $[0, 1) = \{x \in \mathbb{R} \mid 0 \leq x < 1\}$. The following lemma is well known (See for example Ref. 28, pp38–39):

Theorem 3.1 (Kronecker). *Let* $(\theta_1, ..., \theta_r) \in \mathbb{R}^r$ *and* $r \geq 1$. *Then the following statements are equivalent:*

(1) $1, \theta_1, ..., \theta_r$ *are linearly independent over* $\mathbb{Q}$.

(2) $\{(\overline{n\theta_1}, ..., \overline{n\theta_r}) \mid n \in \mathbb{Z}\}$ *is dense in* $[0, 1)^r$.

Using Kronecker's Theorem, we can make the following corollary.

Lemma 3.1. *Suppose that* $\theta \in \mathbb{R} \setminus \mathbb{Q}$ *and* $\delta > 0$. *Then for any* $0 < \epsilon < \delta$, *there exists a sequence* $(m_k, n_k) \in \mathbb{Z}^2$ *such that* $0 < m_k + n_k\theta < \delta \; \forall \; k$, *and* $\lim_{k \to \infty} m_k + n_k\theta = \epsilon$.

Proof. By Theorem 3.1, there is a sequence $\{(M_k, N_k) \in \mathbb{Z}^2\}$ with $\lim_{k \to +\infty} M_k + N_k\theta = 0 \,(\text{mod } \mathbb{Z})$. Take any sequence $\{\epsilon_k\}$ such that $0 < \epsilon_k < \min\{\epsilon, \delta - \epsilon\}$ and $\lim_{k \to +\infty} \epsilon_k = 0$.

Fix some k and take any $p, q_1, q_2 \in \mathbb{N}$ such that

$$\epsilon - \epsilon_k < \frac{q_1}{p} < \frac{q_2}{p} < \epsilon + \epsilon_k < \delta.$$

Take any $M_l, N_l \in \mathbb{Z}$ such that $0 < M_l + N_l\theta < \epsilon - \epsilon_k$ and some $M_j, N_j \in \mathbb{Z}$ such that $0 < M_j + N_j\theta < 1/p$. Then there is some $i \in \mathbb{N}$ such that

$$\epsilon - \epsilon_k < \frac{q_1}{p} \leq M_l + N_l\theta + i(M_j + N_j\theta) < \frac{q_2}{p} < \epsilon + \epsilon_k.$$

Denote $m_k = M_l + iM_j$ and $n_k = N_l + iN_j$, then $0 < \epsilon - \epsilon_k < m_k + n_k\theta < \epsilon + \epsilon_k < \delta$. Thus we have a sequence $\{(m_k, n_k)\}$ such that $0 < m_k + n_k\theta < \delta$ and $\lim_{k \to +\infty} m_k + n_k\theta = \epsilon$. $\square$

For any $m \in I\!N$, we denote $m!! = m! \times (m-1)! \times ... \times 2! \times 1!$ with the convention $0!! = 1$. The following lemma is taken from Ref. 22:

Lemma 3.2. *Let $a_1, a_2, ..., a_m$ be elements of a field, $s_1, s_2, ..., s_m \in I\!N$ with $s_1 + ... + s_m = s$. Consider the following sequence of s exp-polynomial functions in one integer variable: $f_1(n) = a_1^n, f_2(n) = na_1^n, ..., f_{s_1}(n) = n^{s_1-1}a_1^n, f_{s_1+1}(n) = a_2^n, ..., f_{s_1+s_2}(n) = n^{s_2-1}a_2^n, ..., f_s(n) = n^{s_m-1}a_m^n$. Let $V = (v_{pk})$ be the square $s \times s$ matrix where $v_{pk} = f_k(p-1), p, k = 1, ..., s$. Then*

$$\det(V) = \prod_{j=1}^{m} (s_j - 1)!! a_j^{s_j(s_j-1)/2} \prod_{1 \leq i < j \leq m} (a_j - a_i)^{s_i s_j}.$$

Now we will deduce the critical lemmas on generic exp-polynomial functions on $\mathbb{Z}^n$. We first consider the case $n = 2$.

Let

$$f(m,n) = \sum_{(\beta,\gamma) \in B \times \Gamma} a_{\gamma,\beta} m^{\gamma(1)} n^{\gamma(2)} \beta(1)^m \beta(2)^n \qquad (4)$$

be a nonzero generic exp-polynomial function on $\mathbb{Z}^2$, where $\gamma \in \mathbb{Z}_+^2, \beta \in (\mathbb{C}^*)^2$ and Γ, B are finite subsets of $\mathbb{Z}_+^2$ and $(\mathbb{C}^*)^2$ respectively.

Lemma 3.3. *Let a, b be two complex numbers, which are not both roots of unity. Then $G(a,b) = \{(m,n) \in \mathbb{Z}^2 \mid a^m b^n = 1\} = \mathbb{Z}(m_1, n_1)$ for some $m_1, n_1 \in \mathbb{Z}$.*

Proof. Without loss of generality, we may suppose that b is not a root of unity. Choose two different nonzero elements $(m_1, n_1), (m_2, n_2)$ in G with $|m_1|$ being the smallest. Clearly $m_1 \neq 0$. There exist $m_0 \in \mathbb{Z}$ such that $0 \leq m_0 < |m_1|$ and $m_2 = km_1 + m_0$ for some $k \in \mathbb{Z}$. Thus we get $a^{m_0} b^{n_2-kn_1} = 1$. By the choice of m_1, we must have $m_0 = 0$. Then $n_2 - kn_1 = 0$, since b is not a root of unity. Thus $(m_2, n_2) = k(m_1, n_1)$. $\qquad\square$

Lemma 3.4. *Let f be a nonzero generic exp-polynomial function on $\mathbb{Z}^2$ as defined in Eq. (4). Suppose that $\theta \in I\!R \setminus Q$ and $0 < \delta \in I\!R$. Then the following set is infinite:*

$$\{(m,n) \in \mathbb{Z}^2 \mid 0 < m + n\theta < \delta \text{ and } f(m,n) \neq 0\}.$$

Proof. Suppose the above set is finite. Then we can replace δ by a smaller one so that we may assume the above set is empty and

$$\delta < \min\{m + n\theta > 0 \mid (m,n) \in G(\beta(1)^{-1}\beta'(1), \beta(2)^{-1}\beta'(2)), \forall \beta \neq \beta' \in B\},$$

thanks to Lemma 3.3. Then for any $m, n \in I\!N$ with $0 < m+n\theta < \delta$, we have that $\beta(1)^m \beta(2)^n \neq \beta'(1)^m \beta'(2)^n$ if and only if $(\beta(1), \beta(2)) \neq (\beta'(1), \beta'(2))$.

Let $S_N = \{(m, n) \in \mathbb{Z}^2 \mid 0 < m + n\theta < \delta/N\}$ for any $N \in I\!N$. Next we assume that all $(m, n) \in S_N$ for proper N. Then

$$0 = f(km, kn) = \sum_{(\beta, \gamma) \in B \times \Gamma} a_{\gamma, \beta} m^{\gamma(1)} n^{\gamma(2)} k^{\gamma(1)+\gamma(2)} (\beta(1)^m \beta(2)^n)^k,$$

$$\forall \, 0 < k \leq N.$$

Fix $N > |\gamma \times B_{\max}|$. Then by Lemma 3.2, we deduce that for any fixed $\beta \in B$ and any fixed $p \in I\!N$,

$$\sum_{\gamma \in \Gamma, \gamma(1)+\gamma(2)=p} a_{\gamma, \beta} m^{\gamma(1)} n^{\gamma(2)} = 0, \, \forall \, (m, n) \in S_N.$$

For convenience, we may suppose that $\gamma(1), \gamma(2)$ go through the index set $\{0, 1, ..., t\}$, with $t = |\Gamma|$. Fix an $\beta \in B$ and a $p \in I\!N$. Then for all $(m, n) \in S_N$, we have

$$0 = \sum_{\gamma \in \Gamma, \gamma(1)+\gamma(2)=p} a_{\gamma, \beta} m^{\gamma(1)} n^{\gamma(2)}$$

$$= \sum_{\gamma(1)=0}^{\min\{p,t\}} a_\beta(\gamma(1), p - \gamma(1)) \sum_{j=0}^{\gamma(1)} C_{\gamma(1)}^j (m + n\theta)^j (-\theta)^{\gamma(1)-j} n^{\gamma(1)+\gamma(2)-j}$$

$$= \sum_{j=0}^{t} \left(\sum_{\gamma(1)=j}^{\min\{p,t\}} a_\beta(\gamma(1), p - \gamma(1)) C_{\gamma(1)}^j (-\theta)^{\gamma(1)-j} \right) (m + n\theta)^j n^{p-j}, \quad (5)$$

where $C_m^n = (m!)/(n!(m - n)!)$ for $m \geq n \in I\!N$ and $a_\beta(\gamma(1), \gamma(2)) = a_{\gamma, \beta}$, which are supposed to be zero if $(\gamma, \beta) \notin \Gamma \times B$. We denote $A(\gamma, \beta, j) = \sum_{\gamma(1)=j}^{\min\{p,t\}} a_\beta(\gamma(1), p - \gamma(1)) C_{\gamma(1)}^j (-\theta)^{\gamma(1)-j}$ for convenience.

Now we will prove that $A(\gamma, \beta, j) = 0$ for all $0 \leq j \leq t$ and $\gamma = (\gamma(1), p-\gamma(1)) \in \Gamma$. From Lemma 3.1, there exists a sequence $(m_i, n_i) \in S_N$ such that $\lim_{k \to \infty} m_i + n_i \theta = \epsilon$ for some fixed $0 < \epsilon < \delta/N$. Substituting (m_i, n_i) in Eq. (5) and taking a limit, we get $A(\gamma, \beta, 0) = 0$.

Suppose that $A(\gamma, \beta, j) = 0$ for all $0 \leq j < t_0 \leq t$ and $\gamma = (\gamma(1), p - \gamma(1)) \in \Gamma$. Then the coefficients of n^s for $s > t - t_0$ in Eq. (5) are all zero. Substituting (m_i, n_i) in Eq. (5) and taking a limit, we get $A(\gamma, \beta, 0)\epsilon^{t_0} = 0$ and hence $A(\gamma, \beta, t_0) = 0$. Thus we have prove that

$$\sum_{\gamma(1)=j}^{\min\{p,t\}} a_\beta(\gamma(1), p - \gamma(1)) C_{\gamma(1)}^j (-\theta)^{\gamma(1)-j} = 0, \quad (6)$$

for all $0 \leq j \leq t$ and $\gamma = (\gamma(1), p - \gamma(1)) \in \Gamma$.

Taking $p = j$ in Eq. (6), we get that $a_\beta(j, 0) = 0$, $\forall 0 \leq j \leq t$. Now suppose that $a_\beta(j, s) = 0$, $\forall 0 \leq j \leq t, 0 \leq s \leq t_0 - 1$. Taking $p = j + t_0$ in Eq. (6), we get that $\sum_{\gamma(1)=j}^{\min\{p,t\}} a_\beta(\gamma(1), j + t_0 - \gamma(1)) C_{\gamma(1)}^j (-\theta)^{\gamma(1)-j} = 0$, which together with the induction hypothesis gives $a_\beta(j, t_0) = 0$, $\forall 0 \leq j \leq t$. Thus we have proved that $a_\beta(j, s) = 0$, $\forall 0 \leq j \leq t, 0 \leq s \leq t$. That is, $a_{\gamma,\beta} = 0$, $\forall (\gamma, \beta) \in \Gamma \times B$, contradiction. This proves the lemma. $\qquad\square$

We can easily generalize the above result to the following:

Corollary 3.1. *Let $\{f^{(i)}, i \in I\}$ be a finite set of nonzero generic exp-polynomial functions on $\mathbb{Z}^2$ with $\prod_{i \in I} P_{f_i}$ being a free group. Suppose that $\theta \in \mathbb{R} \setminus \mathbb{Q}$ and $0 < \delta \in \mathbb{R}$. Then the following set is infinite:*

$$\{(m, n) \in \mathbb{Z}^2 \mid 0 < m + n\theta < \delta \text{ and } f^{(i)}(m, n) \neq 0, \ \forall i \in I\}.$$

We generalize Lemma 3.4 to generic exp-polynomial functions on $\mathbb{Z}^n$.

Lemma 3.5. *Suppose f is a nonzero generic exp-polynomial function on $\mathbb{Z}^n$. Let $0 < \delta \in \mathbb{R}$ and $\theta \in \mathbb{R}^n$ such that $\theta(1), \theta(2), ..., \theta(n)$ are linearly independent over $\mathbb{Q}$. Then the following set is infinite:*

$$\left\{\alpha \in \mathbb{Z}^n \mid 0 < \sum_{i=1}^n \alpha(i)\theta(i) < \delta \text{ and } f(\alpha) \neq 0\right\}.$$

Proof. We proceed by induction on n. For $n = 2$, the result is Lemma 3.4. Now suppose that the result is true for $2, 3, ..., n - 1$, and we consider the case n.

Suppose that $f(\alpha) = \sum_{(\gamma,\beta) \in \Gamma \times B} a_{\gamma,\beta} \alpha^\gamma \beta^\alpha$ where $\Gamma \times B \subset \mathbb{Z}_+^n \times (\mathbb{C}^*)^n$ is finite. Denote $\bar{\alpha} = (\alpha(1), \alpha(2), ..., \alpha(n), \alpha(n + 1))$, $\hat{\alpha} = (\alpha(1), \alpha(2), ..., \alpha(n - 2), \alpha(n - 1))$ and $\alpha' = (\alpha(n), \alpha(n + 1))$. We define a new function $F(\bar{\alpha}) = f(\alpha(1) + \alpha(n + 1), \alpha(2), ..., \alpha(n))$, which is a nonzero exp-polynomial function in $\bar{\alpha}$. Clearly, we can write

$$F(\bar{\alpha}) = \sum_{(\gamma',\beta') \in \Gamma' \times B'} h_{\gamma',\beta'}(\hat{\alpha}) \alpha'(1)^{\gamma'(1)} \alpha'(2)^{\gamma'(2)} \beta'(1)^{\alpha'(1)} \beta'(2)^{\alpha'(2)},$$

where Γ' and B' are finite subsets of $\mathbb{Z}_+^2$ and $(\mathbb{C}^*)^2$ respectively, and $h_{\gamma',\beta'}$ are exp-polynomial functions in $\hat{\alpha}$. Note that $\alpha'(1) = \alpha(n)$ and $\alpha'(2) = \alpha(n + 1)$.

It is clear that there exists some $(\gamma_0', \beta_0') \in \Gamma' \times B'$ such that $h_{\gamma_0',\beta_0'}$ is nonzero as an exp-polynomial function in $\hat{\alpha}$. Fix this (γ_0', β_0'). Then by the induction hypothesis, there is some $\xi = (\xi(1), ..., \xi(n-1)) \in \mathbb{Z}^{n-1}$ such that $0 < \sum_{i=1}^{n-1} \xi(i)\theta(i) < \delta/2$ and $h_{\gamma_0', \beta_0'}(\xi) \neq 0$.

Fix this ξ. Then define

$$H(\alpha') = F(\xi, \alpha')$$

$$= \sum_{(\gamma', \beta') \in \Gamma' \times \mathrm{B}'} h_{\gamma', \beta'}(\xi)\alpha'(1)^{\gamma'(1)}\alpha'(2)^{\gamma'(2)}\beta'(1)^{\alpha'(1)}\beta'(2)^{\alpha'(2)},$$

which is a nonzero exp-polynomial function in α'. Again by the induction hypothesis, there exists $\zeta = (\zeta(1), \zeta(2)) \in \mathbb{Z}^2$ such that $H(\zeta) \neq 0$ with $0 < \zeta(2)\theta(1) + \zeta(1)\theta(n) < \delta/2$.

Denote $\epsilon = (\epsilon(1), \epsilon(2), .., \epsilon(n)) = (\xi(1) + \zeta(2), \xi(2), ..., \xi(n-1), \zeta(1))$, then we easily have that

$$0 < \sum_{i=1}^{n} \epsilon(i)\theta(i) = \sum_{i=1}^{n-1} \xi(i)\theta(i) + \zeta(2)\theta(1) + \zeta(1)\theta(n) < \delta.$$

Moreover, we have that $f(\epsilon) = F(\xi, \zeta) = H(\zeta) \neq 0$. The lemma follows. $\square$

Corollary 3.2. *Let* $\{f^{(i)}, i \in I\}$ *be a finite set of nonzero generic exp-polynomial functions on* $\mathbb{Z}^n$ *with* $\prod_{i \in I} P_{f_i}$ *being a free group. Let* $0 < \delta \in \mathbb{R}$ *and* $\theta \in \mathbb{R}^n$ *such that* $\theta(1), \theta(2), ..., \theta(n)$ *are linearly independent over* $\mathbb{Q}$. *Then the following set is infinite:*

$$\{\alpha \in \mathbb{Z}^n \mid 0 < \sum_{i=1}^{n} \alpha(i)\theta(i) < \delta \text{ and } f^{(i)}(\alpha) \neq 0, \forall i \in I\}.$$

4. Verma modules over generic exp-polynomial Lie algebras

In this section, we prove Theorem 2.1. Let us first collect some results on total ordered abelian groups. The following result is due to Ref. 29:

Theorem 4.1. *Suppose that* $\prec$ *is a total order on* $G = \mathbb{Z}^n$.

(1) $(\mathbb{Z}^n, \prec)$ *is completely dense if and only if it is isomorphic to some standard* $\mathbb{Z}(\theta_1, ..., \theta_n)$ *with* $n \geq 2$ *and* $\{\theta_i, i = 1, \cdots, n\}$ *being linearly independent over* $\mathbb{Q}$.

(2) *Any* $(\mathbb{Z}^n, \prec)$ *can be expressed as a product of some standard totally ordered abelian groups. More precisely, we can write*

$$(G, \prec) = (G_1, \prec) \times ... \times (G_t, \prec),$$

where each $(G_i, \prec)$ *is isomorphic to some standard totally ordered abelian group.*

By Corollary 3.2 and Theorem 4.1, we have the following result:

Theorem 4.2. *Suppose that $\prec$ is a completely dense total order on $\mathbb{Z}^n$. Let $\{f^{(i)}, i \in I\}$ be a finite set of nonzero generic exp-polynomial functions on $\mathbb{Z}^n$ with $\prod_{i \in I} P_{f_i}$ being a free group. Given any $0 \prec \delta \in \mathbb{Z}^n$, Then the following set is infinite:*

$$\{\alpha \in \mathbb{Z}^n \,|\, 0 \prec \alpha \prec \delta \text{ and } f^{(i)}(\alpha) \neq 0, \,\forall i \in I\}.$$

Let $\widehat{\mathcal{L}}$ be a generic exp-polynomial Lie algebras and $f(\alpha, \beta)$ be the generic exp-polynomial functions defining $\widehat{\mathcal{L}}$ in Def. 2.3. Write

$$f(\alpha, \beta) = \sum_{a \in A, k \in K} h_{a,k}(\beta) a^\alpha \alpha^k, \tag{7}$$

where A and K are finite subset of $(\mathbb{C}^*)^n$ and $\mathbb{Z}_+^n$ respectively, $h_{a,k}$ are generic exp-polynomial functions in β.

Now we are ready to give the proof of Theorem 2.1.

Proof of Theorem 2.1. For any $\alpha_i, \beta_i \in \mathbb{Z}$, by $(\alpha_1, \cdots \alpha_m) \prec (\beta_1, \cdots \beta_m)$ we mean the lexicographical order, i.e., there exists some $1 \leq s \leq m$ such that $\alpha_s \prec \beta_s$ and $\alpha_t = \beta_t$, $\forall\, t > s$. Let v_0 be the highest weight vector of $M = M(\prec, h, c)$.

(1) Suppose that $\prec$ is completely dense. Set

$$\mathcal{B}(m) = \{L_{-\alpha_1} L_{-\alpha_2} \cdots L_{-\alpha_m} \,|\, \alpha_i \in G_+, \, L_{-\alpha_i} \neq 0, \, \alpha_1 \prec \alpha_2 \prec \ldots \prec \alpha_m\}$$

for $m \in \mathbb{Z}_+$, where $G_+ = \{\alpha \in G \,|\, \alpha \succ 0\}$ (similarly for G_-). Denote $\mathcal{B} = \bigcup_{m \geq 0} \mathcal{B}(m)$. Then $\mathcal{B}$ is a basis of $U(\bigoplus_{\alpha \in G_-} \mathbb{C} L_\alpha)$. We define a term order (well-ordering), denoted also by $\prec$, on $\mathcal{B}$ as follows: Given any $x, y \in \mathcal{B}$ with $x = L_{-\alpha_1} L_{-\alpha_2} \ldots L_{-\alpha_m}$ and $y = L_{-\beta_1} L_{-\beta_2} \ldots L_{-\beta_n}$, we say $x \prec y$ if $m < n$ or if $m = n$ and $(\alpha_1, \cdots \alpha_m) \prec (\beta_1, \cdots \beta_n)$. Any $w \in M(\prec, h, c)$ can be written as $w = a_1 x_1 v_0 + a_2 x_2 v_0 + \cdots + a_n x_n v_0$ where $a_1, \cdots, a_n \in \mathbb{C}$, $x_1 \succ \cdots \succ x_n \in \mathcal{B}$. If $a_1 \neq 0$, the highest term of w is defined as $\mathrm{ht}(w) = a_1 x_1 v_0$. For $m \in \mathbb{N}$, we define

$$V(m) = \sum_{x \in \mathcal{B}(k), k \leq m} \mathbb{C} x v_0.$$

To prove Theorem 2.1(1), it will be enough to show that $U(\widehat{\mathcal{L}}) u_0 = M$ for any fixed weight vector u_0 in $M(\prec, h, c) \setminus \mathbb{C} v_0$. (If u_0 is not a weight vector, the theorem follows from the above case.) We first prove that $L_{-\alpha} v_0 \in U(\widehat{\mathcal{L}}) u_0$ for any $\alpha \in G_+$.

Claim 1. There exists a weight vector $u \in U(\widehat{\mathcal{L}})u_0$ such that $u \equiv azv_0(\bmod V(m-1))$, for some $a \in \mathbb{C}^*, z \in \mathcal{B}(m)$ and $m \in \mathbb{N}$.

Proof of Claim 1. Suppose that $u_0 \in V(m) \setminus V(m-1)$. We write

$$u_0 \equiv \sum_{y \in \mathcal{B}(m)} a_y y v_0 \ (\bmod \ V(m-1))$$

with $\mathrm{ht}(u_0) = xv$, where $x = L_{-\alpha_1} L_{-\alpha_2}...L_{-\alpha_m} \in \mathcal{B}(m)$. Let $X = \{y \in \mathcal{B}(m) \,|\, a_y \neq 0\}$. Then x is the unique maximal element in X.

The exp-polynomial function $f(\alpha_m - \alpha, -\alpha_m)$ in $\alpha \in G$ is a nonzero generic exp-polynomial function since $L_{-\alpha_m} \notin Z(\widehat{\mathcal{L}})$. Denote

$$\delta = \min\{\alpha_1, \alpha_m - \beta_k \,|\, y = L_{-\beta_1} L_{-\beta_2}...L_{-\beta_m} \in X, \beta_k \neq \alpha_m\}.$$

Then by Theorem 4.2, there is some $\epsilon_m \in G$ such that $0 \prec \epsilon_m \prec \delta$, $L_{-\epsilon_m} \neq 0$, $L_{\alpha_m - \epsilon_m} \neq 0$, $f(\alpha_m - \epsilon_m, -\alpha_m) \neq 0$ and $h_{a,k}(-\epsilon_m) \neq 0$ for any nonzero functions $h_{a,k}$ defined in (7).

Then we have that

$$[L_{\alpha_m - \epsilon_m}, L_{-\alpha_m}] = f(\alpha_m - \epsilon_m, -\alpha_m)L_{-\epsilon_m} \neq 0,$$

and that

$$[L_{\alpha_m - \epsilon_m}, L_{-\beta_k}] = f(\alpha_m - \epsilon_m, -\beta_k)L_{\alpha_m - \beta_k - \epsilon_m},$$

where $\alpha_m - \beta_k - \epsilon_m \succ 0$ for any $y = L_{-\beta_1} L_{-\beta_2}...L_{-\beta_m} \in X$, $\beta_k \neq \alpha_m$. Consider $u_1 = L_{\alpha_m - \epsilon_m} u_0$, we can get that $\mathrm{ht}(u_1) \in \mathbb{C} L_{-\epsilon_m} L_{-\alpha_1}...L_{-\alpha_{m-1}} v_0$.

Repeat the above process m times, we get

$$u = u_m \equiv L_{-\epsilon_1} L_{-\epsilon_2}...L_{-\epsilon_m} \ (\bmod \ V(m-1))$$

for some $\epsilon_1 \prec \epsilon_2 \prec ... \prec \epsilon_m$. Claim 1 is proved.

Remark 4.1. It is necessary to require that $h_{a,k}(-\epsilon_i) \neq 0$, $1 \leq i \leq m$ for all nonzero functions $h_{a,k}$ as defined in Eq. (7). Thus $f(\alpha, \epsilon_i)$ are nonzero as functions in $-\epsilon_i$, and have the same degree. We refer the reader to the total order defined in the several paragraphs after Eq. (1). We will use these facts in the proof of the following claim:

Claim 2. There is some $\alpha \in G_+$ such that $0 \neq L_{-\alpha} v_0 \in U(\widehat{\mathcal{L}})u_0$.

Proof of Claim 2. Let $u = b_x xv_0 + \sum_{y \in \mathcal{B}(s), s < m} b_y y v_0$ be as in Claim 1, where $x = L_{-\epsilon_1} L_{-\epsilon_2}...L_{-\epsilon_m}$ and $b_x \neq 0$. Let $Y = \{y \,|\, b_y \neq 0\}$. Suppose that the weight of u is $-\lambda$, that is, $\lambda = \epsilon_1 + \epsilon_2 + \cdots + \epsilon_m$. Take $0 \prec \epsilon \prec \delta = \min\{\beta_1 \,|\, y = L_{-\beta_1} L_{-\beta_2}...L_{-\beta_s}, y \in Y\}$.

By the Lie bracket, we can deduce that

$$L_{\lambda-\epsilon}u = \sum_{y \in Y} b_y f_y(\epsilon) L_{-\epsilon}v_0,$$

where

$$f_y(\epsilon) = f(\lambda-\epsilon, -\beta_1)f(\lambda-\epsilon-\beta_1, -\beta_2)\cdots f(\lambda-\epsilon-\beta_1-\beta_2-\cdots-\beta_{s-1}, -\beta_s),$$

for $y = L_{-\beta_1}L_{-\beta_2}\cdots L_{-\beta_s} \in Y$ and

$$f_x(\epsilon) = f(\lambda-\epsilon, -\epsilon_1)f(\lambda-\epsilon-\epsilon_1, -\epsilon_2)\cdots f(\lambda-\epsilon-\epsilon_1-\epsilon_2-\cdots-\epsilon_{m-1}, -\epsilon_m).$$

It is clear that $f_x(\epsilon)$ is nonzero as a function in ϵ. We want to show that $\sum_{y \in Y} b_y f_y(\epsilon)$ is nonzero as a generic exp-polynomial functions in ϵ. Indeed, the highest degree of $f(\alpha, -\epsilon_i), 1 \le i \le m$ regarded as functions in α, is equal to each other (See Remark 4.1 before this claim), and is no less than that of $f(\alpha, -\beta_j), 1 \le j \le s$.

Each $f(\alpha, -\epsilon_i)$ cannot be a constant as a function in α, otherwise it must be zero as a function in α since $f(-\epsilon_i, -\epsilon_i) = 0$, which contradict the choice of ϵ_i. After changing the order on P (see Sect.2 for the definition for P) if necessary, we may assume that the degree $d \in B \times \Gamma$ of each $f(\alpha, \epsilon_i)$ is such that $d \succ 0$.

So $\deg f_x(\epsilon) = md$ while $\deg f_y(\epsilon) \prec md$. Hence $\sum_{y \in Y} b_y f_y(\epsilon)$ is nonzero as an exp-polynomial function. Then by Theorem 4.2, there is some $0 \prec \epsilon \prec \delta$ such that $\sum_{y \in Y} b_y f_y(\epsilon) \neq 0$ and $L_{\lambda-\epsilon-e_1-\ldots-e_j} \neq 0$ for any $j = 0, 1, \ldots 2, m$. Then $0 \neq \sum_{y \in Y} b_y f_y(\epsilon)L_{-\epsilon}v_0 = L_{\lambda-\epsilon}u \in U(\widehat{\mathcal{L}})u_0$, forcing $L_{-\epsilon}v_0 \in U(\widehat{\mathcal{L}})u_0$. Claim 2 is proved.

Claim 3. $L_{-\beta}v_0 \in U(\widehat{\mathcal{L}})u_0, \ \forall \ 0 \prec \beta$ with $L_{-\beta} \neq 0$.

Proof of Claim 3. We still use the nonzero element $L_{-\epsilon}v_0 \in U(\widehat{\mathcal{L}})u_0 \subset M'$. For any $0 \prec \beta \prec \epsilon$ with $L_{-\beta} \neq 0$. Since $f(\epsilon - \alpha, -\epsilon)f(\epsilon - \beta + \alpha, -\epsilon)f(-\beta + \alpha, -\alpha)$ is a nonzero generic exp-polynomial function in α, there exists $\alpha : 0 \prec \alpha \prec \beta - \alpha \prec \beta \prec \epsilon$ such that $L_{\epsilon-\alpha} \neq 0$, $L_{-\alpha} \neq 0$, $L_{\epsilon-\beta+\alpha} \neq 0$, $L_{-\beta+\alpha} \neq 0$ and $f(\epsilon - \alpha, -\epsilon)f(\epsilon - \beta + \alpha, -\epsilon)f(-\alpha, -\beta + \alpha) \neq 0$. Then $f(\epsilon - \alpha, -\epsilon)L_{-\alpha}v_0 = [L_{\epsilon-\alpha}, L_{-\epsilon}]v_0 \in U(\widehat{\mathcal{L}})u_0$, $f(-\epsilon, \epsilon - \beta + \alpha)L_{-\beta+\alpha}v_0 = [L_{-\epsilon}, L_{\epsilon-\beta+\alpha}]v_0 \in U(\widehat{\mathcal{L}})u_0$, $f(-\alpha, -\beta + \alpha)L_\beta v_0 = [L_{-\alpha}, L_{-\beta+\alpha}]v_0 \in U(\widehat{\mathcal{L}})u_0$, and $0 \neq L_{-\beta}v_0 \in U(\widehat{\mathcal{L}})u_0$. That is, $L_{-\beta}v_0 \in U(\widehat{\mathcal{L}})u_0, \ \forall \ 0 \prec \beta \prec \epsilon$ with $L_{-\beta} \neq 0$.

Since $f(-\epsilon - \alpha, -\epsilon) \neq 0$ as an exp-polynomial function in α, there exists $\alpha : 0 \prec \alpha \prec \epsilon/2$ such that $L_{-\epsilon-\alpha} \neq 0$, and $f(-\epsilon - \alpha, -\epsilon) \neq 0$. Then $f(-\epsilon-\alpha, -\epsilon)L_{-2\epsilon-\alpha}v_0 = [L_{-\epsilon-\alpha}, L_{-\epsilon}]v_0 \in U(\widehat{\mathcal{L}})u_0$. In this manner, for any $\beta \in G_+$ with $L_{-\beta} \neq 0$, there exists $\gamma \succ \beta$ such that $0 \neq L_{-\gamma}v_0 \in U(\widehat{\mathcal{L}})u_0$.

Using the above established result we see that $L_{-\beta}v_0 \in U(\widehat{\mathcal{L}})u_0$, $\forall\ 0 \prec \beta$ with $L_{-\beta} \neq 0$. Claim 3 follows.

Sufficiency. Suppose that $f_{h,c}(\alpha)$ is nonzero as an exp-polynomial function in $\alpha \in G$.

There is some $0 \prec \alpha$ such that $L_{\pm\alpha} \neq 0$, and $f_{h,c}(\alpha) \neq 0$. By Claim 3, we see that $f_{h,c}(\alpha)v_0 = [L_\alpha, L_{-\alpha}]v_0 = L_\alpha L_{-\alpha}v_0 \in U(\widehat{\mathcal{L}})u_0$. Thus $v_0 \in U(\widehat{\mathcal{L}})v_0$. This shows that $M(\prec, h, c)$ is irreducible.

Necessity. If $f_{h,c}(\alpha)$ is zero as an exp-polynomial function in α, from Claim 3 we see that M' is a proper irreducible submodule of $M(\prec, h, c)$.

(2) Suppose that $(G, \prec) \cong (H, \prec) \times (K, \prec)$, with $(H, \prec)$ being standard and nonzero. For any $\alpha, \beta \in G$, we write $\beta \succ \mathbb{Z}\alpha$ if $\beta \succ n\alpha$, $\forall\ n \in \mathbb{Z}$. It is clear that $\beta \succ \mathbb{Z}\alpha$ for any $\alpha \in H$ and $\beta \in G_+ \setminus H_+$. Also we have that $G = (G_+ \setminus H_+) \cup H \cup (G_- \setminus H_-)$.

Denote by $M_H(\prec, h, c)$ the $\widehat{\mathcal{L}}(H)$ submodule of $M(\prec, h, c)$ generated by the highest weight vector v_0. Recall that $\widehat{\mathcal{L}}(S) = \bigoplus_{\alpha \in S} \widehat{\mathcal{L}}_\alpha$ for any $S \subset G$.

Easy computations show that $U(\widehat{\mathcal{L}}(G_+ \setminus H^+))M_H(\prec, h, c) = 0$, and $M_H(\prec, h, c)$ can be viewed as a module over $U(\widehat{\mathcal{L}}(G_+ \setminus H_+ + H)) = U(\widehat{\mathcal{L}}(G_+ + H_-))$. Then we have

$$M(\prec, h, c) = U(\widehat{\mathcal{L}}) \bigotimes_{U(\widehat{\mathcal{L}}(G_+ + H_-))} M_H(\prec, h, c)$$

$$= U(\widehat{\mathcal{L}}(G_- \setminus H_-)) \bigotimes_{U(\widehat{\mathcal{L}}(G_+ + H_-))} M_H(\prec, h, c),$$

where the first equation holds as $\widehat{\mathcal{L}}$ modules and the second holds as vector spaces. Obviously, the irreducibility of $M(\prec, h, c)$ over $\widehat{\mathcal{L}}$ implies the irreducibility of $M_H(\prec, h, c)$ over $\widehat{\mathcal{L}}(H)$.

Now we suppose that $M_H(\prec, h, c)$ is irreducible as an $\widehat{\mathcal{L}}(H)$ module. If $K = 0$, then $G = H$ and the result is trivial. Now suppose that $\mathrm{rank}(K) \geq 1$.

Let u be any nonzero vector in $M(\prec, h, c) \setminus \mathbb{C}v_0$ and denote by u_0 the homogeneous component of u with the smallest weight relative to $\prec$.

Set

$$\mathcal{B}'(m) = \left\{ L_{-\beta_1} \cdots L_{-\beta_{m-s}} L_{-\alpha_1} \cdots L_{-\alpha_s} \ \middle| \ \begin{matrix} \alpha_i \in H_+,\ \beta_j \in G_+ \setminus H_+, \\ \alpha_1 \prec \ldots \prec \alpha_s, \\ \beta_1 \prec \ldots \prec \beta_{m-s}, \\ L_{-\beta_i} \neq 0, L_{-\alpha_j} \neq 0 \end{matrix} \right\}$$

and $\mathcal{B}' = \bigcup_{m \geq 0} \mathcal{B}'(m)$. Then $\mathcal{B}'$ is a basis of $U(\widehat{\mathcal{L}}(G_-))$. We define a total order $\prec$ on $\mathcal{B}'$ as follows: Given any $x, y \in \mathcal{B}'$ with

$$x = L_{-\beta_1} \cdots L_{-\beta_{m-s}} L_{-\alpha_1} \cdots L_{-\alpha_s}$$

and

$$y = L_{-\delta_1} \cdots L_{-\delta_{m-r}} L_{-\gamma_1} \cdots L_{-\gamma_r},$$

we say $x \prec y$ if $m < n$, or if $m = n$ and $(\alpha_1, \cdots \alpha_s, \beta_1, \cdots \beta_{m-s}) \prec (\gamma_1, \cdots \gamma_r, \delta_1, \cdots \delta_{m-r})$. Any $w \in M(\prec, h, c)$ can be uniquely written as $w = a_1 x_1 v_0 + a_2 x_2 v_0 + \cdots + a_n x_n v_0$ where $a_1, \cdots, a_n \in \mathbb{C}$, $x_1 \succ \cdots \succ x_n \in \mathcal{B}'$. If $a_1 \neq 0$, the highest term of w is defined as $\mathrm{ht}(w) = a_1 x_1 v_0$.

We also define

$$V'(m) = \sum_{x \in \mathcal{B}(k)', k \leq m} \mathbb{C} \, x v_0,$$

for any $m \in I\!\!N$. Suppose that $u_0 \in V'(m) \setminus V'(m-1)$. Then

$$u_0 \equiv \sum_{y \in \mathcal{B}'(m)} a_y y v_0 \pmod{V'(m-1)} \quad \text{and} \quad \mathrm{ht}(u_0) = a_x x v_0 \neq 0,$$

where $x = L_{-\beta_1} \cdots L_{-\beta_s} L_{-\alpha_{s+1}} \cdots L_{-\alpha_m} \in \mathcal{B}'$, where $\alpha_i \in H_+$ and $\beta_j \in G_+ \setminus H_+$. Let $X = \{y \in \mathcal{B}'(m) \,|\, a_y \neq 0\}$. Then x is the unique maximal element in X.

Case 1. rank$(H) > 1$.

In this case, $(H, \prec)$ is dense and $(H, \prec) \cong Z(\theta(1), ..., \theta(r))$ for some $r > 1$ and $\theta(1), ..., \theta(r)$ linearly independent over $\mathbb{Q}$, by Theorem 4.1.

Since $f(\beta_s - \alpha, -\beta_s)$ is nonzero (the assumption in the theorem) as an exp-polynomial function in $\alpha \in H$, by Theorem 4.2 there is $0 \prec \alpha_s \prec \alpha_{s+1}$ such that $L_{\beta_s - \alpha_s} \neq 0$, $\alpha_s \prec \beta_s - \beta_j$ for all $\beta_j \neq \beta_s$ and $f(\beta_s - \alpha_s, -\beta_s) \neq 0$. Then

$$L_{\beta_s - \alpha_s} x v_0 \equiv p f(\beta_s - \alpha_s, -\beta_s) L_{-\beta_1} \cdots L_{-\beta_{s-1}}$$

$$L_{-\alpha_s} L_{-\alpha_{s+1}} \cdots L_{-\alpha_m} v_0 \pmod{V(m-1)},$$

where p is the number of β_j with $\beta_j = \beta_s$.

Now consider $u_1 = L_{\beta_s - \alpha_s} u_0$. Easy computations shows that

$$0 \neq \mathrm{ht}(u_1) = \mathrm{ht}(L_{\beta_s - \alpha_s} x v_0) \in \mathbb{C} \, L_{-\beta_1} \cdots L_{-\beta_{s-1}} L_{-\alpha_s} L_{-\alpha_{s+1}} \cdots L_{-\alpha_m} v_0.$$

Repeat the above process s times, and we can get $u_1, u_2, ..., u_s$ such that

$$0 \neq \mathrm{ht}(u_q) \in \mathbb{C} \, L_{-\beta_1} \cdots L_{-\beta_{s-q}} L_{-\alpha_{s-q+1}} \cdots L_{-\alpha_m} v_0,$$

where $q = 1, ..., s, \alpha_i \in H_+, \beta_j \in G_+ \setminus H_+$. In particular, $0 \neq \text{ht}(u_s) \in \mathbb{C} \, L_{-\alpha_1} \cdots L_{-\alpha_m} v_0$. Thus we have $u_s \neq 0$ and $\text{ht}(u_s) \in M_H(\prec, c, h)$. Since u_s is a weight vector, we must have $u_s \in M_H(\prec, h, c) \cap U(\widehat{\mathcal{L}})u_0$.

Since $M_H(\prec, h, c)$ is irreducible as a $U(\widehat{\mathcal{L}}(H))$ module, there exists $X \in U(\widehat{\mathcal{L}}(H))$ such that $Xu = Xu_s = v_0$, yielding $U(\widehat{\mathcal{L}})u = M(\prec, h, c)$, which implies our result in this case.

Case 2. rank$(H) = 1$.

In this case, $(H, \prec) \cong (\mathbb{Z}, <)$ and $H = \mathbb{Z}\alpha_0$ for some $\alpha_0 \succ 0$. Then α_0 is the smallest positive element in H.

Recall that $u_0 \equiv \sum_{y \in \mathcal{B}'(m)} a_y y v_0 \pmod{V'(m-1)}$, $\text{ht}(u_0) = a_x x v_0$ and $X = \{y \in \mathcal{B}'(m) \mid a_y \neq 0\}$. We may write

$$x = L_{-\beta_1} \cdots L_{-\beta_s} L_{-k_{s+1}\alpha_0} \cdots L_{-k_m \alpha_0}, \text{ with } 0 < k_{s+1} \leq ... \leq k_m.$$

Denote

$$X' = \left\{ k'_m \; \middle| \; \begin{array}{l} y = L_{-\beta} L_{-\beta_2} \cdots L_{-\beta_s} L_{-k'_{s+1}\alpha_0} \cdots L_{-k'_m \alpha_0}, \\[2mm] 0 < k'_{s+1} \leq ... \leq k'_m, y \in X \end{array} \right\}.$$

Since $f(\beta_1 - \alpha, -\beta_1)$ is nonzero as an exp-polynomial function in $\alpha \in H = \mathbb{Z}\alpha_0$, we first prove that there exists $k \in \mathbb{N}$ such that $f(\beta_1 - k\alpha_0, -\beta_1) \neq 0$, $L_{\beta_1 - k\alpha_0} \neq 0$, and $k > k'$, $\forall k' \in X'$, which is a special case of the following

Claim 4. Suppose that $f(n)$ is a nonzero generic exp-polynomial function in $n \in \mathbb{N}$, then for any $0 \neq k \in \mathbb{Z}$, there exists $n \in k\mathbb{N}$ such that $f(n) \neq 0$.

Proof of Claim 4. We can write $f(n) = \sum_{(a,b) \in \mathbb{N} \times \mathbb{C}^*} c_{a,b} n^a b^n$. Take any $0 \neq k \in \mathbb{Z}$ and suppose that $f(n) = 0$, $\forall n \in k\mathbb{N}$. That is

$$f(ik) = \sum_{(a,b) \in \mathbb{N} \times \mathbb{C}^*} c_{a,b} k^a i^a (b^k)^i = 0, \; \forall i \in \mathbb{N}.$$

Since f is generic, then $b^k = b'^k \Leftrightarrow b = b'$ for any $c_{a,b} \neq 0$ and $c_{a',b'} \neq 0$. Then Lemma 3.2 gives that $c_{a,b} k^a = 0$ and hence $c_{a,b} = 0$ for any $(a, b) \in \mathbb{N} \times \mathbb{C}^*$, contradiction.

Then

$$L_{\beta_1 - k\alpha_0} x v_0 \equiv p f(\beta_1 - k\alpha_0, -\beta_1) L_{-\beta_2} \cdots L_{-\beta_s}$$

$$L_{-k_{s+1}\alpha_0} \cdots L_{-k_m \alpha_0} L_{-k_{m+1}\alpha_0} v_0 \pmod{V(m-1)},$$

where p is the number of β_j with $\beta_j = \beta_1$ and $k_{m+1} = k$.

Again consider $u_1 = L_{\beta_1 - k\alpha_0} u_0$. Then it is also easy to see that

$$0 \neq \mathrm{ht}(u_1) = \mathrm{ht}(L_{\beta_s - k\alpha_0} x v_0)$$

$$\in \mathbb{C} \, L_{-\beta_2} \cdots L_{-\beta_s} L_{-k_{s+1}\alpha_0} \cdots L_{-k_m\alpha_0} L_{-k_{m+1}\alpha_0} v_0.$$

Repeat the above process s times, and we can get $u_1, u_2, ..., u_s$ such that

$$0 \neq \mathrm{ht}(u_q) \in \mathbb{C} \, L_{-\beta_{q+1}} \cdots L_{-\beta_s} L_{-k_{s+1}\alpha_0} \cdots L_{-k_{m+q}\alpha_0} v_0, \quad \forall \ q = 1, ..., s.$$

In particular, $\mathrm{ht}(u_s) \in \mathbb{C} \, L_{-k_{s+1}\alpha_0} \cdots L_{-k_{m+s}\alpha_0} v_0$.

Similarly as in Case 1, we can also deduce that $U(\widehat{\mathcal{L}})u = M(\prec, h, c)$ at last. This complete the proof of Theorem 2.1. $\qquad \square$

Acknowledgments

K.Z. was partially supported by NSF of China (Grant 10871192) and NSERC.

References

1. Y. Billig, Principal vertex operator representations for toroidal Lie algebras, *J. Math. Phys.* 39(7), 3844–3864 (1998).
2. Y. Billig, category of modules for the full toroidal Lie algebra, *Int. Math. Res. Not.*, 2006, Art. ID 68395, 46pp.
3. S. Berman and Y. Billig, Irreducible representations for toroidal Lie lagebras, *J. Algebra*, Vol. 221(1), 188–231 (1999).
4. S. Berman and J. Szmigielski, Principal realization for the extended affine Lie algebra of type sl_2 with coordinates in a simple quantum torus with two generators, *Recent developments in quantum affine algebras and relate topics, Raleigh, NC, 1998, Contemporary Mathematics, 248, American Mathematical Society, Providence, RI*, 1999, pp. 39–67.
5. S. Eswara Rao, Classification of irreducible integrable modules for toroidal Lie algebras with finite dimensional weight spaces, *J. Algebra*, 277 (2004), no. 1, 318–348.
6. S. Eswara Rao, Iterated loop modules and a filtration for vertex representation of toroidal Lie algebras, *Pacific J. Math.*, 171(2), 511–528 (1995).
7. S. Eswara Rao and K. Zhao, Highest weight irreducible representations of quantum tori, *Mathematical Research Letters*, Vol. 11, Nos. 5–6, 615–628 (2004).
8. Y. Gao, Representations of extended affine Lie algebras coordinated by certain quantum tori, *Compositio Math.*, 123(1), 1–25 (2000).
9. M. Golenishcheva-Kutuzova and D. Lebedev, Vertex operator representation of some some quantum tori Lie algebras, *Comm. Math. Phys.*, 148(2), 403–416 (1992).

10. J. Hu, X. Wang, K. Zhao, Verma modules over generalized Virasoro algebras Vir[G], *J. Pure Appl. Algebra*, 177(1), 61–69 (2003).

11. R. Lu and K. Zhao, Verma modules over quantum torus Lie algebras, *Canadian J. Math.*, Vol. 62(2), 382–399 (2010).

12. R. Lu and K. Zhao, Classification of irreducible weight modules over higher rank Virasoro algebras, *Advances in Math.*, Vol. 201(2), 630–656 (2006).

13. R. Lu and K. Zhao, Classification of irreducible weight modules over the twisted Heisenberg-Virasoro algebras, *Comm. Contemp. Math.*, Vol. 12, No. 2, 1–23 (2010).

14. V. Mazorchuk, Classification of simple Harish-Chandra modules over Q-Virasoro algebra, *Math. Nachr.*, 209, 171–177 (2000).

15. V. Mazorchuck, Verma modules over generalized Witt algebras, *Compositio. Math.* 115(1), 21–35 (1999).

16. J. Patera and H. Zassenhaus, The higher rank Virasoro algebras, *Comm. Math. Phys.*, 136, 1–14 (1991).

17. G. Shen, Graded modules of graded Lie algebras of Cartan type. I. Mixed products of modules, *Sci. Sinica, Ser. A*, 29(6), 570–581 (1986).

18. Y. Su, Classification of Harish-Chandra modules over higher rank Virasoro algebras, *Comm. Math. Phys.*, 240, 539–551 (2003).

19. R. Shen, J. Jiang, Y. Su, Verma modules over the generalized Heisenberg-Virasoro algebra, *Comm. Algebra*, 36 (2008), no. 4, 1464–1473.

20. R. V. Moody and A. Pianzola, Lie algebras with triangular decompositions, *Canad. Math. Soc., Ser. Mono. Adv. Texts,* A Wiley-Interscience Publication, John Wiley & Sons Inc., New York, 1995.

21. X. Wang, K. Zhao, Verma modules over Virasoro-like algebras, *J. Australian Math. Soc.*, Vol. 80, no. 2, 179–191 (2006).

22. Y. Billig and K. Zhao, Weight modules over exp-polynomial Lie algebras, *J. Pure Appl. Algebra*, Vol. 191, 23–42 (2004).

23. B. Wilson, Imaginary highest-weight representation theory and symmetric functions, *Comm. Algebra*, Vol. 37 (2009), no. 10, 3729–3749.

24. P. Batra, X. Guo, R. Lu and K. Zhao, Highest weight modules over the pre-exp-polynomial algebras, *Journal of Algebra*, Vol.322, 4163–4180 (2009).

25. A. A. Albert and M. S. Frank, Simple Lie algebras of characteristic p, *Univ. e Politec. Torino Rend. Sem. Mat.* 14, 117–139 (1954–55).

26. D. Z. Dokovic and K. Zhao, Derivations, isomorphisms and second cohomology of generalized Block algebras, *Algebra Colloq.*, 3, 245–272 (1996).

27. R. Block, On torsion-free abelian groups and Lie algebras, *Proc. Amer. Math. Soc.* 9, 613–620 (1958).

28. T. Brocker and T. Dieck, Representations of Compact Lie Group, *Graduate Texts in Mathematics, 98*, Springer-Verlag, 1985.

29. V. M. Kopytov and N. Ya. Medvedev, The theory of lattice-ordered groups, *Mathematics and its Applications, 307.* Kluwer Academic Publishers Group, Dordrecht, 1994, xvi+400 pp.

A FORMAL INFINITE DIMENSIONAL CAUCHY PROBLEM AND ITS RELATION TO INTEGRABLE HIERARCHIES

G. F. HELMINCK

Korteweg-de Vries Institute,
University of Amsterdam,
Science Park 904,
1098 XH Amsterdam, The Netherlands
E-mail: g.f.helminck@uva.nl

E. A. PANASENKO* and A. O. SERGEEVA[†]

Tambov State University,
Internatsionalnaya 33,
392622 Tambov, Russia
* E-mail: panlena_t@mail.ru
[†] E-mail: alena-sr21@vandex.ru

In this paper it is shown under mild assumptions that the local solvability of an infinite dimensional formal Cauchy problem is equivalent to a set of zero curvature relations. The role this type of Cauchy problems plays in integrable systems is illustrated at the hand of lower triangular Toda hierarchies.

Keywords: Infinite dimensional Cauchy problem, zero curvature equations, lower triangular $\mathbb{Z} \times \mathbb{Z}$-matrices, Lax equations, Toda-type hierarchies, linearization, wave matrices

1. Introduction

One considers for infinite size matrices depending formally of an infinite number of parameters the solvability of the analogue of the finite-dimensional local Cauchy problem. If all finite products of the matrices that form the coefficients of the formal power series involved exist , one shows that this solvability is equivalent to a set of zero curvature equations. A natural setting where this type of problem occurs is the construction of solutions of integrable hierarchies. As an illustration we show their role for some lower triangular hierarchies. The content of the various sections is as follows: first we recall the results from the finite dimensional setting. Next we treat the infinite dimensional Cauchy problem and the last section describes the role these Cauchy problems play in the theory of some Toda-type hierarchies.

2. The finite dimensional setting

First one has a look at the scalar case. Consider on an open subset U in $\mathbb{C}^r$ with local coordinates $z_1, \cdots, z_r$ around the point $x_0 \in U$ and a collection of holomorphic functions $\{g_1, \cdots, g_r\}$ on U. It is well-known that the system of differential equations

$$\frac{\partial}{\partial z_i}(f) = g_i f, 1 \leq i \leq r, \tag{1}$$

has for a fixed value $f(x_0) \in \mathbb{C}^r$ locally a unique solution around x_0 if and only if all the compatibility equations

$$\frac{\partial}{\partial z_i}(g_j) = \frac{\partial}{\partial z_j}(g_i), i \text{ and } j \in \{1, \cdots, r\}, \tag{2}$$

hold. Next one considers the vector case of a collection of constant $n \times n$-matrices $\{C_1, \cdots, C_r\}$ and the set of differential equations for a $\mathbb{C}^n$-valued holomorphic function f

$$\frac{\partial}{\partial z_i}(f) = C_i f, 1 \leq i \leq r. \tag{3}$$

If one integrates this equation one variable at a time one sees that a solution, if it exists, must be given by

$$\exp(z_1 C_1) \cdots \exp(z_r C_r)(f(x_0)) = \exp(z_r C_r) \cdots \exp(z_1 C_1)(f(x_0)),$$

where the exponential factors can also be placed in any other order. This last fact implies that all the matrices $\{C_i\}$ have to commute:

$$[C_i, C_j] = 0, \text{ for all } i \text{ and } j \in \{1, \cdots, r\}. \tag{4}$$

The next step is to consider the equation (3) for matrices $\{C_i\}$ that are no longer constant and depend on a neighbourhood of x_0 in a holomorphic way of the local coordinates $z_1, \cdots, z_r$. The conditions under which one has a unique solution of these equations are a mixture of those in (2) and (4). To see how the conditions look like, assume that one has locally for each $\alpha \in \mathbb{C}^n$ a unique solution. Let the $\{e_i\}$ be the standard basis of the $\mathbb{C}^n$. Then there is in a neighbourhood of zero a unique solution f_i of (3) such that $f_i(x_0) = e_i$. Let F be the matrix with the $\{f_i\}$ as columns. It is a holomorphic matrix-valued function with $F(x_0) = \text{Id}$. Hence it is invertible around zero and there holds:

$$\frac{\partial}{\partial z_i}(F) = C_i F, 1 \leq i \leq r. \tag{5}$$

From the equations

$$\frac{\partial}{\partial z_j}\frac{\partial}{\partial z_i}(F) = \frac{\partial}{\partial z_i}\frac{\partial}{\partial z_j}(F)$$

one deduces

$$C_iC_jF + \frac{\partial}{\partial z_j}(C_i)F = C_jC_iF + \frac{\partial}{\partial z_i}(C_j)F.$$

This is equivalent with

$$\left([C_i, C_j] - (\frac{\partial}{\partial z_i}(C_j) - \frac{\partial}{\partial z_j}(C_i))\right)F = 0$$

and since F is invertible, one gets

$$[C_i, C_j] - \left(\frac{\partial}{\partial z_i}(C_j) - \frac{\partial}{\partial z_j}(C_i)\right) = 0. \tag{6}$$

These conditions are necessary, but also sufficient. This result goes back to Cauchy and Kovalevskaya, see[5]

Theorem 2.1. *Let x_0 be a point in $\mathbb{C}^r$ with local coordinates $z_1, \cdots, z_r$ and let the $C_1, \cdots, C_r$ be holomorphic matrix-valued functions on a neighbourhood of x_0. Then the equations*

$$\frac{\partial}{\partial z_i}(f) = C_i f, 1 \le i \le r, \tag{7}$$

possess a unique solution for fixed $f(x_0)$ if and only if the equations (6) hold.

The equations (6) are also called zero curvature relations because they relate to the vanishing of the curvature in the following setting: let Ω be the holomorphic matrix differential 1-form on $U \subset \mathbb{C}^r$ defined by

$$\Omega = \sum_{i=1}^{r} C_i dz_i.$$

The theorem (2.1) gives a necessary and sufficient condition for the solvability of the linear Pfaffian system

$$dy = \Omega y, \qquad y(z) \in \mathbb{C}^n. \tag{8}$$

Consider over U the trivial vector bundle $E = U \times \mathbb{C}^n$. Let the $\{e_i\}$ be the standard basis of $\mathbb{C}^n$. They determine the trivializing sections $\{s_i \mid 1 \le i \le n\}$ of the bundle E by

$$s_i(x) = (x, e_i), x \in U.$$

The matrix Ω of 1-forms over U defines w.r.t. the $\{s_i\}$ a connection $\nabla = d - \Omega$ on the space of sections of E. This connection ∇ is a map from the space $A^0(E)$ of holomorphic sections of E or zero order forms with values in E to the space $A^1(E)$ of 1-forms with values in E satisfying

$$\nabla(f\sigma) = \sigma df + f\nabla(\sigma), \qquad (9)$$

where $\sigma \in A^0(E)$ and $f \in A^0$, the space of holomorphic functions on the manifold. The connection has a natural extension $\nabla : A^p(E) \mapsto A^{p+1}(E)$ from the p-forms $A^p(E)$ with values in E to the $p+1$-forms $A^{p+1}(E)$ with values in E by the formula

$$\nabla(f\sigma) = \sigma \wedge df + f\nabla(\sigma),$$

with f as above and $\sigma \in A^p(E)$. The space of horizontal sections, i.e. those $y = \sum_{i=1}^{p} y_i s_i$ with $\nabla(y) = 0$, is n-dimensional, see,[4] if and only if the curvature

$$\nabla \circ \nabla = -d\Omega + (-\Omega) \wedge (-\Omega)$$

of ∇ is zero, which is equivalent to the Pfaffian system being integrable in the sense of Frobenius, see.[7] The connection ∇ is then also called *integrable*. In terms of the matrices $\{C_i\}$ the integrability of ∇ amounts to the so-called *zero curvature* equations.

3. The Cauchy problem: infinite dimensional case

Here we want to discuss a formal power series version of the Cauchy problem discussed in the foregoing subsection, where both the size of the matrices and the number of variables is infinite. As for the first type of infinity: since both the semi-infinite case of $\mathbb{N} \times \mathbb{N}$-matrices as the bi-infinite case of the $\mathbb{Z} \times \mathbb{Z}$-matrices are important, see respectively[1] and,[?] we present here a joint set-up. In the case that the size is $\mathbb{N} \times \mathbb{N}$, then the product of such a matrix with a vector from $\mathbb{C}^{\mathbb{N}}$ with only a finite number of nonzero coordinates, is well-defined and if one considers $\mathbb{Z} \times \mathbb{Z}$-matrices then one can say the same of the product of a $\mathbb{Z} \times \mathbb{Z}$-matrix with the same type of vectors in $\mathbb{C}^{\mathbb{Z}}$. Let e_j be the vector with its j-th coordinate equal to one and all others equal to zero. Then the $\{e_j \mid j \in \mathbb{N}\}$ are a basis of the finite vectors $\mathbb{C}^{\mathbb{N}}_{fin}$ in $\mathbb{C}^{\mathbb{N}}$ and similarly the $\{e_j \mid j \in \mathbb{Z}\}$ are a basis of the finite vectors $\mathbb{C}^{\mathbb{Z}}$. The variables in the present set-up are the

$$\{z_i \mid i \in I\}, \text{ for some finite or countable index set } I = \{i_1. \cdots, i_n, \cdots\}.$$

One will use a multi-index notation for monomials in these variables: take any $\alpha = (\alpha_i) \in \mathbb{Z}_{\geq 0}^I$ with only a finite number of α_i nonzero. Then one writes

$$z^\alpha := \prod_{i \in I} z_i^{\alpha_i}.$$

On these multi-indices one uses the order relation

$$\alpha \leq \beta \Leftrightarrow \alpha_i \leq \beta_i \text{ for all } i \in I.$$

and the inequality $\alpha < \beta$ means that for one index $\alpha_i < \beta_i$. For simplicity the zero index is denoted by 0. The degree $\deg(\alpha)$ of the multi-index α is given by

$$\deg(\alpha) := \sum_{i \in I} \alpha_i.$$

Assume one has for each $i \in I$ an infinite matrix C_i ($\mathbb{N} \times \mathbb{N}$ or $\mathbb{Z} \times \mathbb{Z}$) with coefficients from the formal power series in the variables $\{z_i\}$, i.e.

$$C_i = \sum_{0 \leq \alpha} C_i(\alpha) z^\alpha, \tag{10}$$

where each $C_i(\alpha)$ is a well-defined complex matrix of the appropriate size. Now one looks for formal power series solutions of the system of equations

$$\frac{\partial}{\partial z_i}(f) = C_i f, i \in I, \tag{11}$$

where $f \in \mathbb{C}[[z_i]]^{\mathbb{N}}$ resp. $\mathbb{C}[[z_i]]^{\mathbb{Z}}$ i.e.

$$f = \sum_{\alpha \geq 0} f(\alpha) z^\alpha, f(\alpha) \in \mathbb{C}^{\mathbb{N}} \text{ resp. } \mathbb{C}^{\mathbb{Z}}.$$

In particular, one needs that all the products $C_i f$ are well-defined vectors in $\mathbb{C}[[z_i]]^{\mathbb{N}}$ resp. $\mathbb{C}[[z_i]]^{\mathbb{Z}}$ and that is the case if all the $C_i(\alpha) f(\beta)$ are well-defined vectors, for then the right hand side of equation (11) becomes

$$\sum_{\gamma \geq 0} \left(\sum_{0 \leq \alpha \leq \gamma} C_i(\alpha) f(\gamma - \alpha) \right) z^\gamma.$$

Let for each $i \in I$, let the multi-index $1(i)$ be defined by

$$1(i)_j = \begin{cases} 0 \text{ for } i \neq j \\ 1 \text{ for } j = i \end{cases}$$

Then the left hand side of equation (11) for the i-th case equals

$$\sum_{\alpha \geq 1(i)} \alpha_i f(\alpha) z^{\alpha - 1(i)}$$

and this leads for all $\gamma \geq 0$ to the identities

$$(\gamma_i + 1) f(\gamma + 1(i)) = \sum_{0 \leq \alpha \leq \gamma} C_i(\alpha) f(\gamma - \alpha). \tag{12}$$

In particular, all the coefficients of f corresponding to degree one multi-indices have to satisfy

$$f(1(i)) = C_i(0) f(0)$$

and with the recursion (12) one shows with induction on the degree of the multi-index α that each coefficient $f(\alpha)$ is a polynomial expression in the matrices $\{C_i(\beta) \mid i \in I, \beta \geq 0\}$ acting on the vector $f(0)$. This proofs that, if f exists, it is uniquely determined by $f(0)$. Thus one sees that the condition on the existence of the product of all the $C_i(\alpha) f(\beta)$ is fulfilled if all finite products of the matrices $\{C_i(\beta) \mid i \in I, \beta \geq 0\}$ exist. This last condition is also necessary to discuss zero curvature relations, so it is natural to make it a general

Assumption 3.1. *All finite products of the* $\{C_i(\beta) \mid i \in I, \beta \geq 0\}$ *are well-defined.*

The assumption (3.1) is clearly satisfied if each $C_i(\beta)$ determines an endomorphism of $\mathbb{C}^{\mathbb{N}}_{fin}$ resp. $\mathbb{C}^{\mathbb{Z}}_{fin}$. Assuming that one has for each element e_j in the basis a solution f_j such that $f_j(0) = e_j$, then one can form the matrix F with the f_j as its columns. It decomposes w.r.t. the variables $\{z_i\}$ as

$$F = F(z) = F(z_{i_1}, z_{i_2}, \cdots) = \sum_{\alpha \geq 0} F(\alpha) z^\alpha = \mathrm{Id} + \sum_{\alpha > 0} F(\alpha) z^\alpha.$$

The matrix F is the formal *fundamental matrix* of the Cauchy problem (11) with $F(0) = \mathrm{Id}$. It satisfies for all $i \in I$

$$\frac{\partial}{\partial z_i}(F) = C_i F. \tag{13}$$

In particular, being able to solve the equations (13) is equivalent to having a unique solution of (11) for each $\beta \in \mathbb{C}^{\mathbb{N}}_{fin}$ resp. $\mathbb{C}^{\mathbb{Z}}_{fin}$. Note that the equations (13) imply that the coefficients $F(\alpha)$ are polynomial expressions in the matrices $C_i(\beta)$ and therefore all finite products of the coefficients $F(\alpha)$ exist. This fact allows you to proof another property that F shares with the fundamental matrix from the finite-dimensional context

Lemma 3.1. *The matrix F has a formal inverse, i.e. there exists a*

$$G = \sum_{\beta \geq 0} G(\beta) z^\beta = \mathrm{Id} + \sum_{\beta > 0} G(\beta) z^\beta$$

such that $FG = \mathrm{Id}$.

Proof. Multiplying the two formal series F en G yields for each $\gamma > 0$:

$$\sum_{0 \leq \alpha \leq \gamma} F(\alpha) G(\gamma - \alpha) = 0 \Leftrightarrow G(\gamma) = - \sum_{0 < \alpha \leq \gamma} F(\alpha) G(\gamma - \alpha).$$

For the multi-indices α of degree 1 this gives you

$$G(\alpha) = -F(\alpha)$$

and by induction on the degree of the relevant multi-indices one shows that each $G(\alpha)$ with $\deg(\alpha) = k$ is a polynomial expression of degree utmost k in the coefficients $\{F(\beta) \mid \deg(\beta) \leq k\}$. $\qquad\square$

Thanks to assumption (3.1) it makes sense to consider the equations

$$\frac{\partial}{\partial z_j} \frac{\partial}{\partial z_i}(F) = \frac{\partial}{\partial z_i} \frac{\partial}{\partial z_j}(F)$$

and they yield

$$C_i C_j F + \frac{\partial}{\partial z_j}(C_i) F = C_j C_i F + \frac{\partial}{\partial z_i}(C_j) F.$$

This is the same as

$$\left([C_i, C_j] - \left(\frac{\partial}{\partial z_i}(C_j) - \frac{\partial}{\partial z_j}(C_i) \right) \right) F = 0$$

and since F is invertible, one gets the formal version of the zero curvature relations

$$[C_i, C_j] - \left(\frac{\partial}{\partial z_i}(C_j) - \frac{\partial}{\partial z_j}(C_i) \right) = 0, \text{ for all } i \text{ and } j \in I. \qquad (14)$$

We have seen now that under the assumption (3.1) the equations (14) are necessary to solve the system (11). They are also sufficient as we will show now. What one needs to show is that the power series coefficients $F(\alpha)$ of F are defined in an unambiguous way by the relations: for all $\gamma \geq 0$

$$(\gamma_i + 1) F(\gamma + 1(i)) = \sum_{0 \leq \alpha \leq \gamma} C_i(\alpha) F(\gamma - \alpha). \qquad (15)$$

First we have a look at the multi-indices α in which only one variable occurs. Then one gets the recursion

$$(k_i + 1)F((k_i + 1)1(i)) = \sum_{0 \le m \le k_i} C_i(m1(i))F((k_i - m)1(i)) \qquad (16)$$

and this determines the $F(k_i 1(i))$ inductively in an unambiguous way starting from $F(0) = \mathrm{Id}$. Moreover the powerseries $F(\cdots, 0, z_i, 0, \cdots)$ is built such that it satisfies

$$\frac{\partial}{\partial z_i}(F(\cdots, 0, z_i, 0, \cdots)) = C_i(\cdots, 0, z_i, 0, \cdots)F(\cdots, 0, z_i, 0, \cdots).$$

The idea is now to use induction w.r.t. n, the number of variables really present in the multi-index α, i.e. the number of elements in $\{i \mid \alpha_i > 0\}$. Thus we may assume that any coefficient $F(\beta)$ corresponding to n or less variables is well-defined and one considers a coefficient $F(\alpha)$ in which $n+1$ variables occur. One may assume that

$$\alpha = \sum_{k=1}^{n+1} \alpha_{i_k} 1(i_k),$$

otherwise one can reduce to this case by rearranging. It is sufficient to show that the compatibility conditions allow you to introduce a well-defined power series

$$F(z_{i_1}, \cdots, z_{i_{n+1}}, 0, \cdots)$$

that satisfies for all $k, 1 \le k \le n+1$,

$$\frac{\partial}{\partial z_{i_k}}(F(z_{i_1}, \cdots, z_{i_{n+1}}, 0, \cdots))$$
$$= C_{i_k}(z_{i_1}, \cdots, z_{i_{n+1}}, 0, \cdots)F(z_{i_1}, \cdots, z_{i_{n+1}}, 0, \cdots).$$

To define the coefficients $F(\alpha), \alpha_{i_k} > 0$ for all $k, 1 \le k \le n+1$, one has $n+1$ possibilities each corresponding to the variable z_{i_k} for which one uses the recursion

$$(\alpha_{i_k})F(\alpha) = \sum_{0 \le \gamma \le \alpha - 1(i_k)} C_{i_k}(\gamma)F(\alpha - 1(i_k) - \gamma). \qquad (17)$$

Each choice gives you a power series $F_k(z_{i_1}, \cdots, z_{i_{n+1}}, 0, \cdots)$ Note that each F_k is constructed in such a way that it satisfies for the variable z_{i_k} the equation

$$\frac{\partial}{\partial z_{i_k}}(F_k) = C_{i_k}(z_{i_1}, \cdots, z_{i_{n+1}}, 0, \cdots)F_k. \qquad (18)$$

Next we show that all the F_k are equal to F_{n+1}. Note that the two series agree on terms with n variables or less so that they agree in particular in the terms with only one variable and hence it is sufficient to prove the equality

$$\frac{\partial}{\partial z_{i_{n+1}}}\frac{\partial}{\partial z_{i_k}}(F_k) - \frac{\partial}{\partial z_{i_k}}\frac{\partial}{\partial z_{i_{n+1}}}(F_{n+1}) = 0.$$

To the left hand side one applies the equations (18) and substitutes the zero curvature relations and this gives for the left hand side

$$\frac{\partial}{\partial z_{i_{n+1}}}(C_{i_k})F_k + C_{i_k}\frac{\partial}{\partial z_{i_{n+1}}}(F_k) - \frac{\partial}{\partial z_{i_k}}(C_{i_{n+1}})F_k - C_{i_{n+1}}\frac{\partial}{\partial z_{i_k}}(F_{n+1}) =$$

$$\frac{\partial}{\partial z_{i_{n+1}}}(C_{i_k})F_{n+1} + C_{i_k}\frac{\partial}{\partial z_{i_{n+1}}}(F_{n+1}) + \frac{\partial}{\partial z_{i_{n+1}}}(C_{i_k})(F_k - F_{n+1}) +$$

$$C_{i_k}\frac{\partial}{\partial z_{i_{n+1}}}(F_k - F_{n+1}) - \frac{\partial}{\partial z_{i_k}}(C_{i_{n+1}})F_{n+1} - C_{i_{n+1}}\frac{\partial}{\partial z_{i_k}}(F_k) -$$

$$\frac{\partial}{\partial z_{i_k}}(C_{i_{n+1}})(F_k - F_{n+1}) - C_{i_{n+1}}\frac{\partial}{\partial z_{i_k}}(F_{n+1} - F_k) = \frac{\partial}{\partial z_{i_{n+1}}}(C_{i_k})F_{n+1} +$$

$$C_{i_k}C_{i_{n+1}}F_{n+1} + \frac{\partial}{\partial z_{i_{n+1}}}(C_{i_k})(F_k - F_{n+1}) + C_{i_k}\frac{\partial}{\partial z_{i_{n+1}}}(F_k - F_{n+1}) -$$

$$\frac{\partial}{\partial z_{i_k}}(C_{i_{n+1}})F_{n+1} - C_{i_{n+1}}C_{i_k}F_k - \frac{\partial}{\partial z_{i_k}}(C_{i_{n+1}})(F_k - F_{n+1}) -$$

$$C_{i_{n+1}}\frac{\partial}{\partial z_{i_k}}(F_{n+1} - F_k) = \frac{\partial}{\partial z_{i_{n+1}}}(C_{i_k})(F_k - F_{n+1}) +$$

$$C_{i_k}\frac{\partial}{\partial z_{i_{n+1}}}(F_k - F_{n+1}) - \frac{\partial}{\partial z_{i_k}}(C_{i_{n+1}})(F_k - F_{n+1}) -$$

$$C_{i_{n+1}}\frac{\partial}{\partial z_{i_k}}(F_{n+1} - F_k).$$

Assume that $F_k - F_{n+1}$ is nonzero and has a non trivial coefficient for the multi-index γ, then $\frac{\partial}{\partial z_{i_{n+1}}}\frac{\partial}{\partial z_{i_k}}(F_k) - \frac{\partial}{\partial z_{i_k}}\frac{\partial}{\partial z_{i_{n+1}}}(F_{n+1})$ has a non trivial coefficient for the multi-index $\gamma - 1(i_k) - 1(i_{n+1})$. If one chooses γ to be one of the lowest mult-indices of $F_k - F_{n+1}$ for which there is a nonzero coefficient, then this contradicts the fact that $F_k - F_{n+1}$, $\frac{\partial}{\partial z_{i_k}}(F_{n+1} - F_k)$ and $\frac{\partial}{\partial z_{i_{n+1}}}(F_{n+1} - F_k)$ do not possess such coefficients. Hence $F_k - F_{n+1}$ has to be zero. We resume this result in a

Theorem 3.1. *Under assumption (3.1) the equations (13) possess a unique formal power series solution F with $F(0) = \mathrm{Id}$ if and only if the zero curvature relations (14) hold for the formal power series $\{C_i\}$.*

One can consider a variant of the equations (13) with the starting value

$F(0) = G$, where G is an invertible matrix different from the identity and in that case one has the same theorem if one makes another assumption

Assumption 3.2. *All finite products of the $C_i(\alpha)$ multiplied from the right with the matrix G give you well-defined matrices.*

This condition is superfluous if all the $C_i(\alpha)$ are finite band matrices, which is the case in the example treated in the next section.

Remark 3.1. In this section we have shown under mild assumptions the existence of a fundamental matrix for a collection of $\mathbb{Z} \times \mathbb{Z}$-matrices $\{C_i \mid i \in I\}$ with coefficients in $\mathbb{C}[[z_i]]$ if the zero curvature equations hold for this collection of matrices. This fundamental matrix was moreover unique up to multiplication from the right with an invertible matrix from $M_{\mathbb{Z}}(R)$. A natural generalization would be to allow singularities in the set-up and to consider localizations $R = S^{-1}\mathbb{C}[[z_i]]$ of the ring $\mathbb{C}[[z_i]]$ with S a multiplicative subset of $\mathbb{C}[[z_i]]$. All the partial derivatives $\frac{\partial}{\partial z_i}$ are then derivations of R. Assuming then that all the matrices $\{C_i \mid i \in I\}$ have coefficients in this bigger algebra R and that all finite products of the $\{C_i\}$ exist , then one might wonder if there exist an invertible matrix $F \in M_{\mathbb{Z}}(R)$ or $M_{\mathbb{N}}(R)$ that satisfies the equations (13). Such a matrix F will be called a *fundamental matrix* corresponding to the $\{C_i \mid i \in I\}$. As in the case of $R = \mathbb{C}[[z_i]]$ a necessary condition are the zero curvature relations (14) for the $\{C_i\}$.

A solution to this problem does not have to exist. Consider namely the case of one variable z_1, the multiplicative set $S = \{z_1^m \mid m \in \mathbb{N}\}$ and the function $C_1(z_1) = \frac{a}{z_1}, a \notin \mathbb{Z}$. Then one verifies directly that the equation

$$\frac{\partial}{\partial z_1}(F(z_1)) = \frac{a}{z_1}F(z_1)$$

does not have a solution of the form $F(z_1) = \sum_{i=m}^{\infty} a_i z_1^i, m \in \mathbb{Z}$.

As for uniqueness, let F_1 and F_2 be fundamental matrices for the $\{C_i\}$ and let $D = F_1^{-1}F_2$ be a well-defined matrix, the analogue of the assumption (3.2) in the present setting. Then there holds

$$\frac{\partial}{\partial z_i}(F_2) = \frac{\partial}{\partial z_i}(F_1)D + F_1\frac{\partial}{\partial z_i}(D) = C_iF_2 + F_1\frac{\partial}{\partial z_i}(D) = C_iF_2 \qquad (19)$$

and thus the matrix D is constant, i.e. $\frac{\partial}{\partial z_i}(D) = 0$ for all i. Despite the lack of a general existence theorem in this localization setting, one can often use geometry to construct fundamental matrices.

4. The Cauchy problem in integrable hierarchies

We want to illustrate in this section how the infinite dimensional Cauchy problem occurs in the theory of integrable hierarchies. We have chosen to demonstrate this at the hand of a lower triangular hierarchy of $\mathbb{Z} \times \mathbb{Z}$-matrices, but a similar role can be shown in the $\mathbb{N} \times \mathbb{N}$-case and for other hierarchies.

4.1. *Lower triangular matrices*

Let R be a commutative ring. Then one writes $M_{\mathbb{Z}}(R)$ for the R-module of $\mathbb{Z} \times \mathbb{Z}$-matrices with coefficients from R. On this space one uses the ordering of columns and rows that is compatible with the finite dimensional case. There are a number of special elements in $M_{\mathbb{Z}}(R)$ that will be used frequently. First of all, there is the basic matrix $E_{(i,j)}$, i and $j \in \mathbb{Z}$, given by

$$(E_{(i,j)})_{mn} = \delta_{im}\delta_{jn}, \tag{20}$$

where δ_{ts}, for t and $s \in \mathbb{Z}$, denotes the Kronecker symbol that is equal to one if $t = s$ and zero otherwise. Thus one can describe every $A = (\alpha_{ij}) \in M_{\mathbb{Z}}(R)$ as a formal linear combination of the basic matrices

$$A = \sum_{i \in \mathbb{Z}} \sum_{j \in \mathbb{Z}} \alpha_{ij} E_{(i,j)} = \sum_{i \in \mathbb{Z}} \sum_{j \in \mathbb{Z}} \alpha_{(i,j)} E_{(i,j)}. \tag{21}$$

The notation $\alpha_{(i,j)}$ for the matrixcoefficient α_{ij} of A will be used only if confusion in the labeling might occur. This will be done without further mentioning. Any map $\Delta : R \to R$ extends in a natural way to a map $\Delta : M_{\mathbb{Z}}(R) \to M_{\mathbb{Z}}(R)$ by putting

$$\Delta(A)_{ij} = \Delta(\alpha_{ij}), \text{ if } A = (\alpha_{ij}).$$

An important role is played by the shift matrix Λ and its powers $\{\Lambda^m \mid m \in \mathbb{Z}\}$ defined by

$$\Lambda^m = \sum_{i \in \mathbb{Z}} E_{(i-m,i)}, m \in \mathbb{Z}.$$

It permits you to decompose each matrix $A \in M_{\mathbb{Z}}(R)$ in diagonals that are handy at explicit computations.

For each nonzero $k \in \mathbb{N}$, one denotes the ring of $k \times k$-matrices with coefficients from the ring R by $M_k(R)$. Assume one has a collection of

$k \times k$-matrices $\{d(ks)|s \in \mathbb{Z}\}$ in $M_k(R)$. To such a collection one associates a diagonal of $k \times k$-blocks $\mathrm{diag}(d(ks))$ in $M_\mathbb{Z}(R)$ given by

$$\mathrm{diag}(d(ks)) := \sum_{s\in\mathbb{Z}}\sum_{\alpha=1}^{k}\sum_{\beta=1}^{k} d(ks)_{\alpha\beta} E_{(sk+\alpha-1, sk+\beta-1)}. \tag{22}$$

The form of this matrix is as follows

$$\begin{pmatrix} \ddots & & \ddots & & \ddots & & \ddots & & \ddots \\ \ddots & \mathbf{d(kn-k)} & & 0 & & 0 & & \ddots \\ \ddots & & 0 & & \mathbf{d(kn)} & & 0 & & \ddots \\ \ddots & & 0 & & 0 & & \mathbf{d(kn+k)} & & \ddots \\ \ddots & & \ddots & & \ddots & & \ddots & & \ddots \end{pmatrix} \tag{23}$$

and justifies the terminology used. For any such $k \geq 1$ one denotes the ring of $k \times k$-block diagonal matrices in $M_\mathbb{Z}(R)$ by

$$\mathcal{D}_k(R) = \{d = \mathrm{diag}(d(ks))|d(ks) \in M_k(R) \text{ for all } s \in \mathbb{Z}\}.$$

One has a diagonal embedding i_k from $M_k(R)$ into $\mathcal{D}_k(R)$ by taking for any $A \in M_k(R)$ all diagonal blocks of $i_k(A)$ equal to A. The elements Λ^{km}, $m \in \mathbb{Z}$, act on $\mathcal{D}_k(R)$ according to the formula

$$\Lambda^{km}\mathrm{diag}(d(ks))\Lambda^{-km} = \mathrm{diag}(d(ks+km)). \tag{24}$$

It implies e.g. that the image of i_k consists of all matrices in $\mathcal{D}_k(R)$ that commute with Λ^k. This brings one to the decomposition in diagonals mentioned above. For, if $A = (\alpha_{ij}) \in M_\mathbb{Z}(R)$, then one puts

Definition 4.1. The *j-th $k \times k$-block diagonal* of any matrix A, $j \in \mathbb{Z}$, is the matrix

$$\sum_{i\in\mathbb{Z}}\sum_{\gamma=1}^{k}\sum_{\beta=1}^{k} \alpha_{(ki-kj+\gamma-1, ki+\beta-1)} E_{(ki-kj+\gamma-1, ki+\beta-1)}.$$

From equation (24) it is clear that the j-th $k \times k$-block diagonal of a $\mathbb{Z} \times \mathbb{Z}$-matrix A can uniquely be written in the form $\mathrm{diag}(d(ks))\Lambda^{kj}$ where $\mathrm{diag}(d(ks))$ belongs to $\mathcal{D}_k(R)$. Thus each matrix $A = (\alpha_{(i,j)}) \in M_\mathbb{Z}(R)$ can uniquely be written as a formal infinite sum

$$A = \sum_{j\in\mathbb{Z}} d_j \Lambda^{kj}, \tag{25}$$

with all the d_j in $\mathcal{D}_k(R)$. A matrix $B \in M_{\mathbb{Z}}(R)$ is called a *finite band* matrix if it has only a finite number of nonzero diagonals. Thanks to the decomposition (25) and the equation (24) it is clear now that multiplying any matrix A from the left or right with a finite band matrix B is well-defined and easily computable. In particular one sees that any matrix that commutes with Λ^k has the form (25) with d_j in the image of i_k. The elements of $\mathcal{D}_k(R)$ were given in terms of $k \times k$-blocks. It will be convenient to work with a similar, compatible decomposition for general matrices $A \in M_{\mathbb{Z}}(R)$. One has then

$$A = \begin{pmatrix} \ddots & & \ddots & & \ddots & & \ddots & & \ddots \\ \ddots & \mathbf{A_{n-1\ n-1}} & A_{n-1\ n} & A_{n-1\ n+1} & \ddots \\ \ddots & A_{n\ n-1} & \mathbf{A_{n\ n}} & A_{n\ n+1} & \ddots \\ \ddots & A_{n+1\ n-1} & A_{n+1\ n} & \mathbf{A_{n+1\ n+1}} & \ddots \\ \ddots & & \ddots & & \ddots & & \ddots & & \ddots \end{pmatrix}, \tag{26}$$

where all the $\{(A)_{ij} := A_{ij} \mid i, j \in \mathbb{Z}\}$ belong to $M_k(R)$ and the (t, s)-entry of A_{ij} is given by $\alpha_{(ik+t-1,jk+s-1)}$. To distinguish between the two notations one will use greek letters for the matrix coefficients of A in R and latin ones for the coefficients in $M_k(R)$. E.g. for the matrices $i_k(F), F \in M_k(R)$ and Λ^k their $k \times k$-block decomposition looks respectively like

$$i_k(F) = \begin{pmatrix} \ddots & \ddots & \ddots & \ddots & \ddots \\ \ddots & \mathbf{F} & 0 & 0 & \ddots \\ \ddots & 0 & \mathbf{F} & 0 & \ddots \\ \ddots & 0 & 0 & \mathbf{F} & \ddots \\ \ddots & \ddots & \ddots & \ddots & \ddots \end{pmatrix}, \quad \Lambda^k = \begin{pmatrix} \ddots & \ddots & \ddots & \ddots & \ddots \\ \ddots & \mathbf{0} & \text{Id} & 0 & \ddots \\ \ddots & 0 & \mathbf{0} & \text{Id} & \ddots \\ \ddots & 0 & 0 & \mathbf{0} & \ddots \\ \ddots & \ddots & \ddots & \ddots & \ddots \end{pmatrix}.$$

The basic matrices can be multiplied and satisfy

$$E_{(i,j)} E_{(m,k)} = \delta_{jm} E_{(i,k)}.$$

In general, for arbitrary matrices $A = (\alpha_{ij}) = (A_{ij})$ and $B = (\beta_{ij}) = (B_{ij})$ in $M_{\mathbb{Z}}(R)$, one cannot define the product AB. However, if it exists, its ts-block is given by

$$(AB)_{ts} = \sum_{j \in \mathbb{Z}} A_{tj} B_{js}. \tag{27}$$

In particular, this expression makes sense if both A and B belong to either of the classes of upper or lower triangular matrices that will be introduced now.

Definition 4.2. An element A in $M_{\mathbb{Z}}(R)$ is called *upper $k \times k$-block triangular of level m*, if it can be written as

$$A = \sum_{j \geq m} a_j \Lambda^{kj}, \quad \text{with } a_j \in \mathcal{D}_k(R). \qquad (28)$$

One calls m the *order* of A in Λ^k, if a_m is nonzero.

The collection of all upper $k \times k$-block triangular elements of level m in Λ^k one denotes by $UT_{\geq m}(\Lambda^k)$ or simply $UT_{\geq m}$. For the set of all upper $k \times k$-block triangular matrices one uses the notations

$$UT(R) := \bigcup_{m \in \mathbb{Z}} UT_{\geq m}(\Lambda^k) =: UT.$$

In the same spirit one has the notations

$$UT_{<m}(\Lambda^k) := \{A \mid A \in UT, A = \sum_{j<m} a_j \Lambda^{kj}, a_j \in \mathcal{D}_k(R)\} =: UT_{<m},$$

$UT_{\leq m}(\Lambda^k)$ and $UT_{>m}(\Lambda^k)$. Clearly $UT(R)$ is independent of k and one speaks also simply of *upper triangular matrices* if the size of the diagonals needs less emphasis. It is a direct verification that UT with the product (27) forms an R-algebra.

Similarly one can introduce the collection of lower triangular matrices $LT(R)$ as the collection of matrices in $M_{\mathbb{Z}}(R)$ for which all the j-th diagonals are zero if j is sufficiently large. For completeness sake one introduces analogous terminology and notations for this class of matrices.

Definition 4.3. An element A in $M_{\mathbb{Z}}(R)$ is called *lower $k \times k$-block triangular of level m*, if it can be written as

$$A = \sum_{j \leq m} d_j \Lambda^{kj}, \quad \text{with } d_j \in \mathcal{D}_k(R).$$

One calls m the *order* of A in Λ^k, if d_m is nonzero.

The collection of all lower $k \times k$-block triangular elements of level m one denotes by $LT_{\leq m}(\Lambda^k)$. The notations $LT_{>m}(\Lambda^k), LT_{<m}(\Lambda^k)$ and $LT_{\geq m}(\Lambda^k)$ speak for themselves. For the set of all lower $k \times k$-block triangular matrices one uses the notations

$$LT(R) := \bigcup_{m \in \mathbb{Z}} LT_{\leq m}(\Lambda^k) =: LT.$$

Clearly $LT(R)$ is independent of k and one speaks also simply of *lower triangular matrices* if the size of the diagonals needs less emphasis. It is a direct verification that LT with the product (27) forms an R-algebra.

Any matrix $A = \sum_{j\in\mathbb{Z}} d_j \Lambda^{kj}$ as in (25) splits into a component $A_{\geq 0} \in UT_{\geq 0}(\Lambda^k)$ and a component $A_{<0} \in LT_{<0}(\Lambda^k)$ according to

$$A_{\geq 0} := \sum_{j\geq 0} d_j \Lambda^{kj} \text{ and } A_{<0} = \sum_{j<0} d_j \Lambda^{kj}. \tag{29}$$

The decomposition of $M_{\mathbb{Z}}(R)$ into the direct sum $UT_{\geq 0}(\Lambda^k) \oplus LT_{<0}(\Lambda^k)$ of two Lie algebras is basic for the construction of solutions for the integrable hierarchy to be introduced in the next subsection.

4.2. *The Lax equations of the* $(\Lambda^k, \mathbf{h}_{\geq 0})$*-hierarchy*

The starting point of the hierarchy in the title of this subsection is the choice of a basic set of commuting directions in the Lie algebra $UT_{\geq 0}(\Lambda^k)$. Let $\mathbf{h}$ be a maximal commutative subalgebra of $M_k(\mathbb{C})$ with the basis $\{E_\alpha \mid \alpha \in \{1, \cdots, m_0\}\}$. Then the basic commuting directions are the

$$\{\Lambda^{ki} i_k(E_\alpha) \mid i \geq 0, \alpha \in \{1, \cdots, m_0\}\}.$$

The flow parameter w.r.t. the direction $\Lambda^{ki} i_k(E_\alpha)$ is $t_{i\alpha}$ and for the ring R we choose from now on the formal power series $\mathbb{C}[[t_{i\alpha}]]$ in all these parameters. The $\{t_{i\alpha}\}$ will play the role of the variables $\{z_i \mid i \in I\}$ from the second section. One considers now perturbations $\mathcal{L}$ and $\mathcal{U}_\alpha$ in $M_{\mathbb{Z}}(R)$ of the basic directions Λ^k and $i_k(E_\alpha)$. These perturbations should have the form

$$\mathcal{L} = \Lambda^k + \sum_{i\leq 0} l_i \Lambda^{ki} \text{ resp. } \mathcal{U}_\alpha = i_k(E_\alpha) + \sum_{j<0} u_{j,\alpha} \Lambda^{kj}. \tag{30}$$

The matrix $P_{i\alpha} := \mathcal{L}^i \mathcal{U}_\alpha$ is considered then as a perturbation of the basic direction $\Lambda^{ki} i_k(E_\alpha)$. The first property that one requires of $\mathcal{L}$ and the $\mathcal{U}_\alpha, \alpha \in \{1, \cdots, m_0\}$, is that they preserve the commutativity

$$[\mathcal{L}, \mathcal{U}_\alpha] = 0 \text{ and } [\mathcal{U}_\alpha, \mathcal{U}_\beta] = 0 \text{ for all } \alpha \text{ and } \beta. \tag{31}$$

This is trivially satisfied if $\mathcal{L}$ and the $\mathcal{U}_\alpha$ are obtained by dressing the basic directions, i.e.

$$\mathcal{L} = U\Lambda^k U^{-1} \text{ and } \mathcal{U}_\alpha = U i_k(E_\alpha) U^{-1},$$

with a matrix $U = \text{Id} + \sum_{i<0} u_i \Lambda^{ki}$. More crucial, are the evolution equations w.r.t. the parameters $\{t_{i\alpha}\}$ that the perturbations should satisfy. If we define for each $i \geq 0$ and each $\alpha \in \{1, \cdots, m_0\}$, the matrix $B_{i\alpha} = (P_{i\alpha})_{\geq 0}$, then these equations read

$$\partial_{i\alpha}(\mathcal{L}) = [B_{i\alpha}, \mathcal{L}] \text{ and } \partial_{i\alpha}(\mathcal{U}_\beta) = [B_{i\alpha}, \mathcal{U}_\beta], \tag{32}$$

where $\partial_{i\alpha}$ is a short hand notation for the partial derivative w.r.t. $t_{i\alpha}$. The equations (31) and (32) form the equations of the $(\Lambda^k, \mathbf{h}_{\geq 0})$-hierarchy and the evolution equations (32) are called the Lax equations of the hierarchy, since their form is similar to the operator form of the Korteweg de Vries equation found by Peter Lax, see.[6] The basic directions Λ^k and the $i_k(E_\alpha)$ are a trivial solution of these equations.

4.3. The zero curvature form of the hierarchy

By applying an isomorphism between $LT(R)$ and $UT(R)$ one can deduce from[3] that the following holds

Proposition 4.1. *Let $\mathcal{L}$ and the $\mathcal{U}_\alpha$ be perturbations in $M_{\mathbb{Z}}(R)$ of the form (30) that satisfy the commutation relations (31). Then the Lax equations for $\mathcal{L}$ and the $\mathcal{U}_\alpha$ are equivalent to the zero curvature relations for the operators $\{B_{i\alpha}\}$ in $M_{\mathbb{Z}}(R)$: for all n and $m \geq 0$ and all β and α in $\{1, \cdots, m_0\}$*

$$\partial_{n\alpha}(B_{m\beta}) - \partial_{m\beta}(B_{n\alpha}) - [B_{n\alpha}, B_{m\beta}] = 0. \tag{33}$$

Note that the coefficients of the power series $B_{m\beta}$ are finite band matrices in $UT_{\geq 0}(\Lambda^k)$. Hence the conditions in the assumptions (3.1) and (3.2) hold so that there is for each invertible matrix G a unique fundamental matrix F_G of the Cauchy problem associated with the $\{B_{i\alpha}\}$ such that $F_G(0) = G$. If one chooses $G \in UT_{\geq 0}(\Lambda^k)$, then F_G also belongs to $UT_{\geq 0}(\Lambda^k)$. To a solution $\mathcal{L}$ and the $\mathcal{U}_\alpha$ of the $(\Lambda^k, \mathbf{h}_{\geq 0})$-hierarchy one can associate still another set of zero curvature conditions. There holds namely

Lemma 4.1. *Define for all $i \geq 0$ and all α in $\{1, \cdots, m_0\}$ the strict lower triangular matrix $D_{i\alpha} = B_{i\alpha} - P_{i\alpha} = -(P_{i\alpha})_{<0}$. Then the $\{D_{i\alpha}\}$ satisfy the zero curvature relations*

$$\partial_{n\alpha}(D_{m\beta}) - \partial_{m\beta}(D_{n\alpha}) - [D_{n\alpha, m\beta}] = 0. \tag{34}$$

Proof. The proof is a combination of the zero curvature relations for the $\{B_{i\alpha}\}$ and the Lax equations for the $\mathcal{L}$ and the $\mathcal{U}_\alpha$. $\square$

Since $D_{i\alpha} = -(P_{i\alpha})_{<0}$ the coefficients of the power series $D_{i\alpha}$ are matrices in $LT_{<0}(\Lambda^k)$. Therefore the conditions of assumption (3.1) hold and there is a unique fundamental matrix H of the Cauchy problem related to the $\{D_{i\alpha}\}$ such that $H(0) = \mathrm{Id}$. The matrix H belongs then to $LT_{\leq 0}(\Lambda^k)$. In the next subsection we describe its role within the integrable hierarchy.

4.4. *Wave matrices for the* $(\Lambda^k, \mathbf{h}_{\geq 0})$*-hierarchy*

A well-known way to construct solutions of an integrable hierarchy is through finding solutions of its linearization. We will explain the role the two Cauchy problems play here. One starts with the potential solutions, namely operators $\mathcal{L}$ and $\mathcal{U}_\alpha$ in $LT(R)$ of the form (30). The *linearization of the* $(\Lambda^k, \mathbf{h}_{\geq 0})$*-hierarchy* consists of the following equations for the matrices $\mathcal{L}$ and the $\mathcal{U}_\alpha$

$$\mathcal{L}\phi = \phi\Lambda^k, U_\alpha\phi = \phi i_k(E_\alpha) \text{ and } \partial_{i\alpha}(\phi) = B_{i\alpha}\phi, \tag{35}$$

where ϕ is an invertible $\mathbb{Z} \times \mathbb{Z}$-matrix with coefficients from $\mathbb{C}[[t_{i\alpha}]]$ that will be specified later. To get the Lax equations for $\mathcal{L}$ one applies the derivation $\partial_{i\alpha}$ to the first equation in (35) and substitutes the last one. This leads to the following manipulations

$$\partial_{i\alpha}(\mathcal{L}\phi - \phi\Lambda^k) = \partial_{i\alpha}(\mathcal{L})\phi + \mathcal{L}(\partial_{i\alpha}(\phi)) - (\partial_{i\alpha}(\phi))\Lambda^k$$
$$= \partial_{i\alpha}(\mathcal{L})\phi + \mathcal{L}B_{i\alpha}\phi - B_{i\alpha}\phi\Lambda^k$$
$$= \{\partial_{i\alpha}(\mathcal{L}) - [B_{i\alpha}, \mathcal{L}\,]\}\phi = 0. \tag{36}$$

Since ϕ is invertible, this implies the Lax equations for $\mathcal{L}$. For the operator $\mathcal{U}_\beta$ one applies $\partial_{i\alpha}$ to the second equation in (35) and substitutes the last one. Thus one gets

$$\partial_{i\alpha}(\mathcal{U}_\beta\phi - \phi i_k(E_\alpha)) = \partial_{i\alpha}(\mathcal{U}_\beta)\psi + \mathcal{U}_\beta(\partial_{i\alpha}(\phi)) - (\partial_{i\alpha}(\phi))i_k(E_\alpha)$$
$$= \partial_{i\alpha}(\mathcal{U}_\beta)\phi + \mathcal{U}_\beta B_{i\alpha}\phi - B_{i\alpha}\phi i_k(E_\alpha)$$
$$= \{\partial_{i\alpha}(\mathcal{U}_\beta) - [B_{i\alpha}, \mathcal{U}_\beta\,]\}\phi = 0 \tag{37}$$

and the same argument yields the the Lax equations for $\mathcal{U}_\beta$.

Next we discuss the form of the matrices ϕ in the linearization (35). As one can see from the equations one needs a left action of matrices like $\mathcal{L}$, the $\mathcal{U}_\beta$ and all the $B_{i\alpha}$ on the ϕ and a right one for matrices like Λ^k and the $i_k(E_\alpha)$. To realize the first, one builds a left $LT(R)$-module structure on the functions ϕ. The actual form of the elements in the module is guided by the trivial solution $\mathcal{L} = \Lambda^k$ and $\mathcal{U}_\alpha = i_k(E_\alpha)$ of the hierarchy. In that case the equations (35) become

$$\Lambda^k\phi = \phi\Lambda^k, i_k(E_\alpha)\phi = \phi i_k(E_\alpha) \text{ and } \partial_{i\alpha}(\phi) = (\Lambda^{ik} i_k(E_\alpha))\phi. \tag{38}$$

The equations (38) have the following unique solution

$$\phi_\infty := \exp(\sum_{i=0}^{\infty} \sum_{\alpha=1}^{m_0} t_{i\alpha} i_k(E_\alpha)\Lambda^{ik}), \tag{39}$$

satisfying $\phi_\infty(0) = \text{Id}$. The module for the linearization will consist of perturbations in $LT(R)$ of this trivial solution ϕ_∞. Note namely that the coefficients in the power series ϕ_∞ are finite band matrices in $LT(\mathbb{C})_{\geq 0}$. Therefore the product of an element in $LT(R)_{\leq m}$ and the power series ϕ_∞ defines a well-defined formal power series in the $\{t_{i\alpha}\}$ with coefficients from $LT(\mathbb{C})$. Consider therefore the collection $M^{(\infty)}(\Lambda^k)$ consisting of the products

$$\{\sum_{j=-\infty}^{N} d_j \Lambda^{kj}\}\phi_\infty, \text{ where } d_j \in \mathcal{D}_k(R). \tag{40}$$

The elements of $M^{(\infty)}(\Lambda^k)$ are called *oscillating matrices at infinity*, since they are products of a series in Λ^k that has a pole around infinity and the exponential term ϕ_∞, which has as a series in Λ^k an essential singularity at infinity. Clearly there is a well-defined left action of $LT(R)$ on it. Also the right multiplication with Λ^k and $i_k(E_\alpha)$ is well-defined on elements of $M^{(\infty)}(\Lambda^k)$. In particular $M^{(\infty)}(\Lambda^k)$ is a free $LT(R)$-module with generator ϕ_∞. An element $\phi \in M^{(\infty)}(\Lambda^k)$ is invertible as soon as one knows that $\phi = \hat{\phi}\phi_\infty$ with $\hat{\phi} \in LT(R)$ invertible. In this last case the equation $\mathcal{L}\phi = \phi\Lambda^k$ is equivalent with

$$\mathcal{L} = \hat{\phi}\Lambda^k\hat{\phi}^{-1}$$

and the equation $\mathcal{U}_\alpha\phi = \phi i_k(E_\alpha)$ is the same as

$$\mathcal{U}_\alpha = \hat{\phi}i_k(E_\alpha)\hat{\phi}^{-1}.$$

To get matrices $\mathcal{L}$ and the $\mathcal{U}_\alpha$ of the right shape, one gauges the $\hat{\phi}$. An oscillating matrix at infinity $\phi = \hat{\phi}\phi_\infty$, where $\hat{\phi} = \sum_{i \leq 0} d_i \mathcal{L}_0^i$ with $d_0 = \text{Id}$, is called *a wave matrix at infinity* for the matrices $\mathcal{L} = \hat{\phi}\Lambda^k\hat{\phi}^{-1}$ and the $U_\alpha = \hat{\phi}i_k(E_\alpha)\hat{\phi}^{-1}$, if it satisfies the equations (35). Since the manipulations to get the Lax equations are well-defined on such a ϕ, the matrices $\mathcal{L}$ and the $\mathcal{U}_\alpha$ form a solution of the $(\Lambda^k, \mathbf{h}_{\geq 0})$-hierarchy.

4.5. *The relation with Cauchy problems*

Let $\phi = \hat{\phi}\phi_\infty$ be a wave matrix at infinity for the matrices $\mathcal{L} = \hat{\phi}\Lambda^k\hat{\phi}^{-1}$ and $U_\alpha = \hat{\phi}i_k(E_\alpha)\hat{\phi}^{-1}$. Then we know from proposition 4.1 and lemma 4.1 that both the $\{B_{i\alpha}\}$ and the $\{D_{i\alpha} = B_{i\alpha} - \mathcal{L}^i U_\alpha\}$ satisfy the zero curvature relations, which implies the solvability of the related Cauchy problems. The matrix $\hat{\phi}$ is the solution for the Cauchy problem for the $\{D_{i\alpha}\}$, for the equation $\partial_{i\alpha}(\phi) = B_{i\alpha}\phi$ yields:

$$\partial_{i\alpha}(\phi) = \partial_{i\alpha}(\hat{\phi})\phi_\infty + \hat{\phi}\Lambda^{ki}i_k(E_\alpha)\phi_\infty$$
$$= \{\partial_{i\alpha}(\hat{\phi})\hat{\phi}^{-1} + \mathcal{L}^i U_\alpha\}\hat{\phi}\phi_\infty = B_{i\alpha}\phi. \tag{41}$$

This implies the required identity for $\hat{\phi}$

$$\partial_{i\alpha}(\hat{\phi})\hat{\phi}^{-1} + (\mathcal{L}^i U_\alpha)_{<0} = 0 \Leftrightarrow \partial_{i\alpha}(\hat{\phi}) = D_{i\alpha}\hat{\phi}. \tag{42}$$

As we saw above there exists for each invertible $G \in M_{\mathbb{Z}}(\mathbb{C})$ a unique fundamental matrix ψ_G satisfying $\psi_G(0) = G$ and

$$\partial_{i\alpha}(\psi_G) = B_{i\alpha}\psi_G.$$

Note that $\hat{\phi}^{-1}\psi_{\mathrm{Id}}$ is a well-defined power series with coefficients in $LT(\mathbb{C})$. The same holds for $\hat{\phi}^{-1}\psi_G$ if the products of all the coefficients of the power series $\hat{\phi}^{-1}\psi_{\mathrm{Id}}$ with G are well-defined. The evolution of $\hat{\phi}^{-1}\psi_{\mathrm{Id}}$ w.r.t. the $t_{i\alpha}$ takes the form

$$\partial_{i\alpha}(\hat{\phi}^{-1}\psi_{\mathrm{Id}}) = -\hat{\phi}^{-1}\partial_{i\alpha}(\hat{\phi})\hat{\phi}^{-1}\psi_{\mathrm{Id}} + \hat{\phi}^{-1}\partial_{i\alpha}(\psi_{\mathrm{Id}})$$
$$= (\hat{\phi}^{-1}\{-D_{i\alpha} + B_{i\alpha}\}\hat{\phi})\hat{\phi}^{-1}\psi_{\mathrm{Id}} = (\hat{\phi}^{-1}\mathcal{L}^i U_\alpha\hat{\phi})\hat{\phi}^{-1}\psi_{\mathrm{Id}}$$
$$= \Lambda^{ki}i_k(E_\alpha)\hat{\phi}^{-1}\psi_{\mathrm{Id}}. \tag{43}$$

Hence if one considers the well defined product $\phi_\infty^{-1}\hat{\phi}^{-1}\psi_{\mathrm{Id}}$, then its evolution w.r.t. all parameters is constant

$$\partial_{i\alpha}(\phi_\infty^{-1}\hat{\phi}^{-1}\psi_{\mathrm{Id}}) = -\Lambda^{ki}i_k(E_\alpha)(\phi_\infty^{-1}\hat{\phi}^{-1}\psi_{\mathrm{Id}}) + (\phi_\infty^{-1}\Lambda^{ki}i_k(E_\alpha)\hat{\phi}^{-1}\psi_{\mathrm{Id}}) = 0.$$

In other words, $\hat{\phi}^{-1}\psi_{\mathrm{Id}} = \phi_\infty g$, with g a constant matrix. This is the crucial decomposition to construct solutions of the hierarchy.

For, consider reversely an invertible matrix $g \in M_{\mathbb{Z}}(\mathbb{C})$ such that

$$\phi_\infty g = \varphi^{-1}\psi, \text{ with } \varphi - \mathrm{Id} \in LT(R)_{<0} \text{ and } \psi = \sum_{i=0}^{\infty} d_i \Lambda^{ki}, \tag{44}$$

with d_0 invertible. Assume that $\varphi\phi_\infty g = \psi$ is a well-defined matrix then there holds

Theorem 4.1. *Let g, φ and ψ satisfy the conditions above. The matrix $\varphi\phi_\infty$ is the wave matrix at infinity of the operators*

$$\mathcal{L}(\varphi) = \varphi\Lambda^k\varphi^{-1} \text{ and } \mathcal{U}_\alpha(\varphi) = \varphi i_k(E_\alpha)\varphi^{-1}$$

Proof. It suffices to show the last equation of the linearization. Applying $\partial_{i\alpha}$ to the identity $\varphi\phi_\infty = \psi g^{-1}$ yields that

$$
\begin{aligned}
\partial_{i\alpha}(\varphi\phi_\infty) &= \{\partial_{i\alpha}(\varphi)\varphi^{-1} + \varphi\Lambda^{ki}i_k(E_\alpha)\varphi^{-1}\}\varphi\phi_\infty \\
&= \{\partial_{i\alpha}(\varphi)\varphi^{-1} + \mathcal{L}^iU_\alpha\}\varphi\phi_\infty = \{\partial_{i\alpha}(\psi)\psi^{-1}\}\psi g^{-1} \\
&= \{\partial_{i\alpha}(\psi)\psi^{-1}\}\varphi\phi_\infty.
\end{aligned}
\tag{45}
$$

By comparing the different expressions in (45) for the elements in the free $LT(R)$-module $M^{(\infty)}(\Lambda^k)$ and by taking the upper triangular part "≥ 0" one gets the identity

$$
\partial_{i\alpha}(\psi)\psi^{-1} = (\mathcal{L}(\varphi)^iU_\alpha(\varphi))_{\geq 0}
$$

and this proves the statement in the theorem. $\qquad\square$

Remark 4.1. Geometric objects like infinite dimensional homogeneous spaces, see (^{2}), can be used to produce suitable matrices g that satisfy the conditions in (44).

References

1. Adler, M.; van Moerbeke, P.: *Group factorization, moment matrices, and Toda lattices.* Internat. Math. Res. Notices 1997, no. **12**, 555-572.
2. Helminck, G. F.; Opimakh, A. V.: *The zero curvature form of integrable hierarchies in the $\mathbb{Z} \times \mathbb{Z}$-matrices* to appear in Algebra Colloquium.
3. Helminck, A. G.; Helminck, G. F.; Opimakh, A. V.: *The relative frame bundle of an infinite dimensional flag variety and solutions of integrable hierarchies*, Theoretical and Mathematical Physics, **165**(3): 1610-1636 (2010).
4. Kobayashi, S.: *Differential Geometry of Complex Vector Bundles, Publications of the Mathematical Society of Japan* **15** (1987), Iwanami Shoten Publishers and Princeton University Press.
5. Kovalevskaya, S. V.: *Nauchnye raboty*, Izdat. Akad. Nauk., Moscow (1948), 368 pp.
6. Lax, P. D.: *Integrals of nonlinear equations of evolution and solitary waves*, Comm. Pure Appl. Math. **21** (1968), 467-490.
7. Warner, F.: *Foundations of differentiable manifolds and Lie groups*, Springer-Verlag, New York, 1983.

PARTIALLY HARMONIC TENSORS AND QUANTIZED SCHUR–WEYL DUALITY

JUN HU* and ZHANKUI XIAO†

*School of Mathematics and Statistics, University of Sydney, NSW 2006, Australia
Email: junhu303@yahoo.com.cn
†School of Mathematical Sciences, Huaqiao University, Quanzhou, 362021, P.R. China
Email: zhkxiao@gmail.com

Let $l, n \in \mathbb{N}$ and $\widetilde{\mathfrak{g}} \in \{\mathfrak{o}_{2l+1}(\mathbb{C}), \mathfrak{sp}_{2l}(\mathbb{C}), \mathfrak{o}_{2l}(\mathbb{C})\}$. Let V be the natural representation of the quantized enveloping algebra $\mathbb{U}_q(\widetilde{\mathfrak{g}})$ and $\mathfrak{B}_{n,q}$ the specialized Birman–Murakami–Wenzl algebra with parameters depending on $\widetilde{\mathfrak{g}}$ (see (2) and Section 2 for their definitions). There is a Schur–Weyl duality between $\mathbb{U}_q(\widetilde{\mathfrak{g}})$ and $\mathfrak{B}_{n,q}$ on $V^{\otimes n}$. We prove that $V^{\otimes n}$ can be decomposed into a direct sum of the subspaces of q-partially harmonic tensors of different valences and there is also a Schur–Weyl duality between the Hecke algebra $\mathcal{H}_q(\mathfrak{S}_n)$ of type A_{n-1} and $\mathbb{U}_q(\widetilde{\mathfrak{g}})$ on the subspaces of q-harmonic tensors. We construct explicitly a complete set of maximal vectors in $V^{\otimes n}$ and identify some simple $\mathfrak{B}_{n,q}$-modules that they generate with the simple $\mathfrak{B}_{n,q}$-modules arising from Enyang's cellular basis of $\mathfrak{B}_{n,q}$.

Keywords: Birman–Murakami–Wenzl algebra, quantized enveloping algebra, Hecke algebra, q-partially harmonic tensor.

1. Introduction

Let $l, n \in \mathbb{N}$ and $\widetilde{\mathfrak{g}} \in \{\mathfrak{o}_{2l+1}(\mathbb{C}), \mathfrak{sp}_{2l}(\mathbb{C}), \mathfrak{o}_{2l}(\mathbb{C})\}$. Let $X \in \{B, C, D\}$ denote the type of the Cartan datum. Let q be an indeterminate over $\mathbb{Q}$ if $X \in \{C, D\}$; or let $q^{1/2}$ be an indeterminate over $\mathbb{Q}$ if $X = B$. *Throughout this paper*, we set

$$K := \begin{cases} \mathbb{Q}(q), & \text{if } X \in \{C, D\}; \\ \mathbb{Q}(q^{1/2}), & \text{if } X = B. \end{cases},$$

$$N := \begin{cases} 2l, & \text{if } X \in \{C, D\}; \\ 2l + 1, & \text{if } X = B. \end{cases}, \quad m := \begin{cases} 2l + 1, & \text{if } X = B; \\ l, & \text{if } X = C; \\ 2l, & \text{if } X = D. \end{cases} \quad (1)$$

If $X \in \{B, D\}$ then we use $\mathbb{U}_q(\mathfrak{so}_N(\mathbb{C}))$ to denote the quantized enveloping algebra over K associated to the special orthogonal Lie algebra $\mathfrak{so}_N(\mathbb{C})$.

In this case, let θ be the K-algebra automorphism of $\mathbb{U}_q(\mathfrak{so}_N)$ which is defined on generators as follows: if $X = B$ then we define $\theta = \mathrm{id}$; if $X = D$ then we define $\theta(e_i) = e_{\theta(i)}$, $\theta(f_i) = f_{\theta(i)}$, $\theta(k_i) = k_{\theta(i)}$, $i = 1, 2, \cdots, l$, where e_i, f_i, k_i are the Chevalley generators of $\mathbb{U}_q(\mathfrak{so}_N)$ and

$$\theta(i) := i + \delta_{i,l-1} - \delta_{i,l}, \ i = 1, 2, \cdots, l.$$

Let $K[\langle\theta\rangle]$ denote the group algebra of the order 2 cyclic group $\langle\theta\rangle$. Following [15], we define $\mathbb{U}_q(\widetilde{\mathfrak{g}}) := \mathbb{U}_q(\mathfrak{so}_N) \rtimes K[\langle\theta\rangle]$ and call $\mathbb{U}_q(\widetilde{\mathfrak{g}})$ the quantized enveloping algebra associated to the orthogonal Lie algebra $\mathfrak{o}_N(\mathbb{C})$. It is clear that $\mathbb{U}_q(\widetilde{\mathfrak{g}})$ can be endowed with a Hopf algebra structure which extends the Hopf algebra structure on $\mathbb{U}_q(\mathfrak{so}_N)$ and such that $\Delta(\theta) = \theta \otimes \theta$, $\varepsilon(\theta) = 1$, $S(\theta) = \theta$, where Δ, ε, S are the comultiplication, counit and antipode respectively.

If $X = C$ then we use $\mathbb{U}_q(\widetilde{\mathfrak{g}})$ to denote the quantized enveloping algebra associated to the symplectic Lie algebra $\mathfrak{sp}_{2l}(\mathbb{C})$.

Let $\mathfrak{B}_n(r, q)$ be the Birman–Murakami–Wenzl algebra introduced in [1, 22], see 2.1 for precise definition. In this paper, we shall only use some specialized Birman–Murakami–Wenzl algebras. That is

$$\mathfrak{B}_{n,q} := \begin{cases} \mathfrak{B}_n(q^{2l}, q), & \text{if } X = B; \\ \mathfrak{B}_n(-q^{2l+1}, q), & \text{if } X = C; \\ \mathfrak{B}_n(q^{2l-1}, q), & \text{if } X = D. \end{cases} \tag{2}$$

Let V be the natural representation of the quantized enveloping algebra $\mathbb{U}_q(\widetilde{\mathfrak{g}})$ over K. By [15], there is a right action of $\mathfrak{B}_{n,q}$ on $V^{\otimes n}$ which commutes with the left action of $\mathbb{U}_q(\widetilde{\mathfrak{g}})$. We refer the reader to Section 2 for the precise definitions of V and these actions. Let φ, ψ be the natural algebra homomorphisms

$$\varphi : (\mathfrak{B}_{n,q})^{\mathrm{op}} \to \mathrm{End}_{\mathbb{U}_q(\widetilde{\mathfrak{g}})}(V^{\otimes n}), \quad \psi : \mathbb{U}_q(\widetilde{\mathfrak{g}}) \to \mathrm{End}_{\mathfrak{B}_{n,q}}(V^{\otimes n}),$$

respectively. For each integer $k \geq 0$, we write $\lambda \vdash k$ to mean that $\lambda = (\lambda_1, \lambda_2, \dots)$ is a partition of k, and denote by $\ell(\lambda)$ the largest integer i such that $\lambda_i \neq 0$. If $\lambda \vdash k$, then we write $|\lambda| = k$ and denote by $\lambda' = (\lambda'_1, \lambda'_2, \cdots)$ the conjugate of λ. In particular, $\lambda'_1 = \ell(\lambda)$. The following results are referred as the quantized Schur–Weyl duality of types B, C, D.

Theorem 1.1. *([5,10.2],[15]) 1) Both φ and ψ are surjective; if $m \geq n$, then φ is an isomorphism;*

2) there is an irreducible $\mathbb{U}_q(\widetilde{\mathfrak{g}})$-$\mathfrak{B}_{n,q}$-*bimodule decomposition*

$$V^{\otimes n} = \bigoplus_{f=0}^{[n/2]} \bigoplus_{\lambda \in \mathcal{P}_f(X)} \widetilde{L}(\lambda) \otimes \widetilde{D}_{f,\lambda},$$

where for each (f, λ), $\widetilde{L}(\lambda)$ *and* $\widetilde{D}_{f,\lambda}$ *denotes the corresponding irreducible* $\mathbb{U}_q(\widetilde{\mathfrak{g}})$-*module and irreducible* $\mathfrak{B}_{n,q}$-*module respectively (see [15, Proposition 4.2] and Section 3 for details), and*

$$\mathcal{P}_f(X) := \begin{cases} \{\lambda \vdash n - 2f \mid \ell(\lambda) \leq m\}, & \text{if } X = C; \\ \{\lambda \vdash n - 2f \mid \lambda_1' + \lambda_2' \leq m\}, & \text{if } X \in \{B, D\}, \end{cases}$$

When specializing q to 1 (if $X \in \{C, D\}$) or $q^{1/2}$ to 1 (if $X = B$), one recovers the classical Schur–Weyl duality of types B, C, D. Equivalently, one can replace V by an N dimensional symplectic or orthogonal vector space $V_{\mathbb{C}}$ over $\mathbb{C}$, $\mathbb{U}_q(\mathfrak{g})$ by the group algebra $\mathbb{C}G$, where $G \in \{Sp(V), O(V)\}$, and $\mathfrak{B}_{n,q}$ by the specialized Brauer algebra $\mathfrak{B}_n$, where $\mathfrak{B}_n = \mathfrak{B}_n(-N)$ if $X = C$; or $\mathfrak{B}_n = \mathfrak{B}_n(N)$ if $X \in \{B, D\}$. Then there is a Schur–Weyl duality between the group algebra $\mathbb{C}G$ and the specialized Brauer algebra $\mathfrak{B}_n$ on $V_{\mathbb{C}}^{\otimes n}$ (cf. [2–4]). Note that characteristic-free versions of the classical Schur–Weyl duality of types B, C, D are also available by the results in [6–8]. The integral version of the quantized Schur–Weyl duality of type C was proved in [17].

Let $E_1, E_2, \cdots, E_{n-1}$ be the contraction generators of the specialized Birman–Murakami–Wenzl algebra $\mathfrak{B}_{n,q}$, see Definition 2.1. For each integer $0 \leq f \leq [n/2]$, let $\mathfrak{B}_n^{(f)}$ be the two-sided ideal of $\mathfrak{B}_{n,q}$ generated by $E_1 E_3 \cdots E_{2f-1}$. Set

$$\mathcal{H}\mathcal{T}_{f,q}^{\otimes n} := \{v \in V^{\otimes n}\mathfrak{B}_n^{(f)} \mid vx = 0, \forall\, x \in \mathfrak{B}_n^{(f+1)}\}.$$

We call $\mathcal{H}\mathcal{T}_{f,q}^{\otimes n}$ the subspace of *q-partially harmonic tensors of valence* f and $\mathcal{H}_q(V^{\otimes n}) := \mathcal{H}\mathcal{T}_{0,q}^{\otimes n}$ the subspace of *q-harmonic tensors*. We refer the reader to [13, (10.3.1)] for the classical case where q is specialized to 1. Each $\mathcal{H}\mathcal{T}_{f,q}^{\otimes n}$ is indeed a $\mathbb{U}_q(\widetilde{\mathfrak{g}})$-$\mathfrak{B}_{n,q}$-bimodule. The first main result of this paper is the following theorem.

Theorem 1.2. *For each integer g with $0 \leq g \leq [n/2]$, there is a* $\mathbb{U}_q(\widetilde{\mathfrak{g}})$-$\mathfrak{B}_{n,q}$-*bimodule isomorphism:*

$$\mathcal{H}\mathcal{T}_{g,q}^{\otimes n} \cong \bigoplus_{\substack{\lambda \in \mathcal{P}(X) \\ \|\lambda\| = n-2g}} \widetilde{L}(\lambda) \otimes \widetilde{D}_{(n-|\lambda|)/2,\lambda} \cong V^{\otimes n}\mathfrak{B}_n^{(g)}/V^{\otimes n}\mathfrak{B}_n^{(g+1)},$$

where $\|\lambda\|$ is defined as in Definition 3.2 and $\mathcal{P}(X) := \cup_{0 \leq f \leq [n/2]} \mathcal{P}_f(X)$. In particular, we have a $\mathbb{U}_q(\widetilde{\mathfrak{g}})$-$\mathfrak{B}_{n,q}$-bimodule decomposition $V^{\otimes n} = \bigoplus_{f=0}^{[n/2]} \mathcal{H}T_{f,q}^{\otimes n}$.

To prove the above theorem, we shall construct explicitly a complete set of maximal vectors of the form $z_{f,\lambda}, z_{f,\lambda}^{\pm}$ in $V^{\otimes n}$, see Definition 3.2, where f is an integer with $0 \leq f \leq [n/2]$ and $\lambda \in \mathcal{P}_f(X)$. Since $\mathfrak{B}_{n,q}/\mathfrak{B}_n^{(1)} \cong \mathcal{H}_q(\mathfrak{S}_n)$, the Iwahori–Hecke algebra of type A_{n-1}, it follows that the space $\mathcal{H}_q(V^{\otimes n})$ of q-harmonic tensors naturally becomes a $\mathbb{U}_q(\widetilde{\mathfrak{g}})$-$\mathcal{H}_q(\mathfrak{S}_n)$-bimodule. Let φ_0, ψ_0 be the natural algebra homomorphisms

$$\varphi_0 : (\mathcal{H}_q(\mathfrak{S}_n))^{\mathrm{op}} \to \mathrm{End}_{\mathbb{U}_q(\widetilde{\mathfrak{g}})}\big(\mathcal{H}_q(V^{\otimes n})\big),$$

$$\psi_0 : \mathbb{U}_q(\widetilde{\mathfrak{g}}) \to \mathrm{End}_{\mathcal{H}_q(\mathfrak{S}_n)}\big(\mathcal{H}_q(V^{\otimes n})\big)$$

respectively. The second main result of this paper is:

Theorem 1.3. *1) Both φ_0 and ψ_0 are surjective, and if $N - l \geq n$, then φ_0 is an isomorphism;*

2) there is an irreducible $\mathbb{U}_q(\widetilde{\mathfrak{g}})$-$\mathcal{H}_q(\mathfrak{S}_n)$-bimodule decomposition

$$\mathcal{H}_q(V^{\otimes n}) \cong \bigoplus_{\substack{\lambda \vdash n \\ \ell(\lambda) \leq N - l}} \widetilde{L}(\lambda) \otimes S(\lambda),$$

where $S(\lambda)$ denotes the irreducible $\mathcal{H}_q(\mathfrak{S}_n)$-module corresponding to λ.

3) if $N - l < n$, then the kernel of φ_0 is the two-sided ideal of $\mathcal{H}_q(\mathfrak{S}_n)$ generated by $\sum_{\sigma \in \mathfrak{S}_{N-l+1}} (-q)^{-\ell(\sigma)} \widehat{T}_\sigma$, where $\{\widehat{T}_w\}_{w \in \mathfrak{S}_n}$ are the standard basis of $\mathcal{H}_q(\mathfrak{S}_n)$.

Since q is an indeterminate, $\mathfrak{B}_{n,q}$ is a quasi-hereditary cellular algebra and hence each simple module appears as a unique simple head of some cell module. In [11], Enyang has constructed a cellular basis for $\mathfrak{B}_{n,q}$ which was obtained by lifting a Murphy basis of $\mathcal{H}_q(\mathfrak{S}_{n-2f})$ (for each $0 \leq f \leq [n/2]$) to $\mathfrak{B}_{n,q}$. We consider the following Murphy basis of $\mathcal{H}_q(\mathfrak{S}_{n-2f})$:

$$\left\{ \widehat{T}_{d(s)^{-1}} \Big(\sum_{w \in \mathfrak{S}_{\{2f+1,\cdots,n\}}} (-q)^{-\ell(w)} \widehat{T}_w \Big) \widehat{T}_{d(t)} \ \Big| \ s, t \in \mathrm{Std}_f(\lambda) \right\},$$

where $\mathrm{Std}_f(\lambda)$ denotes the set of standard λ-tableaux with entries in $\{2f + 1, \cdots, n\}$. Lifting this Murphy basis we get a version of Enyang's cellular basis for $\mathfrak{B}_{n,q}$. For each $\lambda \vdash n - 2f$ with $0 \leq f \leq [n/2]$, we use $S(f, \lambda)$ to denote the corresponding cell module and $D(f, \lambda)$ the unique simple head of $S(f, \lambda)$. The next theorem is the third main result of this paper (which was used in [19]).

Theorem 1.4. *Let λ be a partition of $n - 2f$ with $0 \leq f \leq [n/2]$ and $\ell(\lambda) \leq N - l$. If $X \in \{B, C\}$, then $z_{f,\lambda} \mathfrak{B}_{n,q} \cong \widetilde{D}_{f,\lambda} \cong D(f, \lambda')$; if $X = D$ and $\ell(\lambda) = l$, then $z_{f,\lambda}^{\pm} \mathfrak{B}_{n,q} \cong \widetilde{D}_{f,\lambda} \cong D(f, \lambda')$.*

Combining this theorem with [15, Proposition 4.2], we get that the dimension of $D(f, \lambda')$ is given by the number of $(-2l)$-admissible up-down λ-tableaux of length n if $X = C$; or the number of N-admissible λ-tableaux of length n if $X \in \{B, D\}$, see [24] and [20, Definition 3.4] for the precise definitions of admissible up-down tableaux.

We set $\mathscr{A}'$ to be $\mathbb{Z}[q, q^{-1}]$ if $X = C$; or $\mathbb{Z}[q, q^{-1}, 1/2]$ if $X = D$; or $\mathbb{Z}[q^{1/2}, q^{-1/2}, 1/2]$ if $X = B$. Then $\mathfrak{B}_{n,q}$ has a natural $\mathscr{A}'$-form $\mathfrak{B}_{n,\mathscr{A}'}$. For any field k which is an $\mathscr{A}'$-algebra, the specialized BMW algebra $\mathfrak{B}_{n,k}$ over k is well–defined. The most difficult part of the paper is the construction of those maximal vectors $z_{f,\lambda}, z_{f,\lambda}^{\pm}$. In particular, when $X \in \{B, D\}$, the possibility that $\ell(\lambda)$ can be bigger than l makes the construction quite tricky. In all our explicit construction, $z_{f,\lambda}, z_{f,\lambda}^{\pm} \in V_{\mathscr{A}'}^{\otimes n}$ and we construct them with the hope that the right ideals $z_{f,\lambda} \mathfrak{B}_{n,q}, z_{f,\lambda}^{\pm} \mathfrak{B}_{n,q}$ are stable under base change in the following sense.

Conjecture 1.1. *Let f be an integer with $0 \leq f \leq [n/2]$ and $\lambda \in \mathcal{P}_f(X)$. Supose that $\ell(\lambda) \leq N - l$ and $X \in \{B, C\}$ or $X = D$ and $\ell(\lambda) < l$. Then $z_{f,\lambda} \mathfrak{B}_{n,\mathscr{A}'}$ is a free $\mathscr{A}'$-module of finite rank. Furthermore, for any field k which is an $\mathscr{A}'$-algebra, the natural map $z_{f,\lambda} \mathfrak{B}_{n,\mathscr{A}'} \otimes_{\mathscr{A}'} k \to z_{f,\lambda} \mathfrak{B}_{n,k}$ is an isomorphism. In particular, $\dim z_{f,\lambda} \mathfrak{B}_{n,k}$ is independent of the ground field k and the specialization of q in k. The same statements apply to $z_{f,\lambda}^{\pm}$ if $X = D$ and $\ell(\lambda) = l$.*

Indeed, the conjecture was known to be true in the case when $X = C$ (in particular, $\ell(\lambda) \leq l$) and q is specialized to 1 by the main result in [18].

2. Quantized Enveloping Algebra and BMW Algebra

In this section we shall collect some basic knowledge about the quantized enveloping algebra $\mathbb{U}_q(\widehat{\mathfrak{g}})$ and the specialized Birman–Murakami–Wenzl algebra $\mathfrak{B}_{n,q}$ as well as their actions on the n-tensor space $V^{\otimes n}$.

Let $(a_{i,j})_{l \times l}$ be the Cartan matrix of type X_l. We set

$$q_i = \begin{cases} q^2, & \text{if } X = B \text{ and } i \neq l, \text{ or } X = C \text{ and } i = l; \\ q, & \text{if } X = B \text{ and } i = l, \text{ or } X = D, \text{ or } X = C \text{ and } i \neq l; \end{cases}$$

$$[k]_i = \frac{q_i^k - q_i^{-k}}{q_i - q_i^{-1}} \text{ and } [k]_i^! = [k]_i [k-1]_i \cdots [1]_i.$$

Let $\mathfrak{g} \in \{\mathfrak{so}_{2l+1}(\mathbb{C}), \mathfrak{sp}_{2l}(\mathbb{C}), \mathfrak{so}_{2l}(\mathbb{C})\}$ be the simple Lie algebra corresponding to X_l. The quantized enveloping algebra $\mathbb{U}_q(\mathfrak{g})$ is the associative unital algebra over K generated by $e_i, f_i, k_i, k_i^{-1}, i = 1, ..., m$, subject to the relations:

$$k_i k_i^{-1} = k_i^{-1} k_i = 1, \qquad k_i^{\pm 1} k_j^{\pm 1} = k_j^{\pm 1} k_i^{\pm 1},$$

$$k_i e_j k_i^{-1} = q^{a_{i,j}} e_j, \qquad k_i f_j k_i^{-1} = q^{-a_{i,j}} f_j,$$

$$[e_i, f_i] = \frac{t_i - t_i^{-1}}{q_i - q_i^{-1}} \delta_{i,j} \quad \text{where} \quad t_i = \begin{cases} k_i^2, & \text{if } X = B \text{ and } i \neq l; \\ k_l, & \text{if } X = B \text{ and } i = l; \\ k_i, & \text{if } X = D \text{ or } X = C \text{ and } i \neq l; \\ k_l^2, & \text{if } X = C \text{ and } i = l. \end{cases}$$

$$\text{if } i \neq j , \quad \sum_{k=0}^{1-a_{i,j}} (-1)^k e_i^{(k)} e_j e_i^{(1-a_{i,j}-k)} = 0,$$

$$\text{if } i \neq j , \quad \sum_{k=0}^{1-a_{i,j}} (-1)^k f_i^{(k)} f_j f_i^{(1-a_{i,j}-k)} = 0,$$

where $e_i^{(k)} = e_i^k/[k]_i!$ and $f_i^{(k)} = f_i^k/[k]_i!$. Note that the notation k_i in [15, 4.7–4.13] corresponds to the notation t_i in this paper.

$\mathbb{U}_q(\mathfrak{g})$ is a Hopf algebra with coproduct Δ, counit ε and antipode S defined on generators by

$$\Delta(e_i) = e_i \otimes 1 + t_i \otimes e_i, \quad \Delta(f_i) = 1 \otimes f_i + f_i \otimes t_i^{-1},$$

$$\Delta(k_i) = k_i \otimes k_i,$$

$$\varepsilon(e_i) = \varepsilon(f_i) = 0, \quad \varepsilon(k_i) = 1,$$

$$S(e_i) = -t_i^{-1} e_i, \quad S(f_i) = -f_i t_i, \quad S(k_i) = k_i^{-1}.$$

For each integer $1 \leq i \leq N$, we define $i' := N + 1 - i$. Let V_K be the K-vector space with basis $v_1, v_2, \cdots, v_N$. It is well known that there is a *natural representation* of $\mathbb{U}_q(\mathfrak{g})$ on V_K. By [15], it can be extended to a *natural representation* of $\mathbb{U}_q(\tilde{\mathfrak{g}})$. We recall the definition as follows. Let $1 \leq i < l$, $j \in \{1, 2, \cdots, N\}$. Then

$$e_i v_j := \begin{cases} v_i, & \text{if } j = i+1, \\ -v_{(i+1)'}, & \text{if } j = i', \\ 0, & \text{otherwise}; \end{cases} \qquad f_i v_j := \begin{cases} v_{i+1}, & \text{if } j = i, \\ -v_{i'}, & \text{if } j = (i+1)', \\ 0, & \text{otherwise}; \end{cases}$$

$$\theta v_j := \begin{cases} -v_j, & \text{if } X = B; \\ v_j, & \text{if } X = D \text{ and } j \notin \{l, l+1\}; \\ v_l, & \text{if } X = D \text{ and } j = l+1; \\ v_{l+1}, & \text{if } X = D \text{ and } j = l. \end{cases}$$

$$k_i v_j := \begin{cases} q v_j, & \text{if } X \neq B \text{ and either } j = i \text{ or } j = (i+1)', \\ q^{-1} v_j, & \text{if } X \neq B \text{ and either } j = i+1 \text{ or } j = i', \\ q^{1/2} v_j, & \text{if } X = B \text{ and either } j = i \text{ or } j = (i+1)', \\ q^{-1/2} v_j, & \text{if } X = B \text{ and either } j = i+1 \text{ or } j = i', \\ v_j, & \text{otherwise}, \end{cases}$$

$$e_l v_j := \begin{cases} q^{1/2} v_l, & \text{if } X = B \text{ and } j = l+1, \\ -v_{l+1}, & \text{if } X = B \text{ and } j = l+2, \\ v_l, & \text{if } X = C \text{ and } j = l+1, \\ v_{l-1}, & \text{if } X = D \text{ and } j = l+1, \\ -v_l, & \text{if } X = D \text{ and } j = l+2, \\ 0, & \text{otherwise}; \end{cases}$$

$$f_l v_j := \begin{cases} q^{-1/2} v_{l+1}, & \text{if } X = B \text{ and } j = l, \\ -v_{l+2}, & \text{if } X = B \text{ and } j = l+1, \\ v_{l+1}, & \text{if } X = C \text{ and } j = l, \\ v_{l+1}, & \text{if } X = D \text{ and } j = l-1, \\ -v_{l+2}, & \text{if } X = D \text{ and } j = l, \\ 0, & \text{otherwise}, \end{cases}$$

$$k_l v_j := \begin{cases} q v_j, & \text{if } X = B \text{ and } j = l, \\ q^{-1} v_j, & \text{if } X = B \text{ and } j = l+2, \\ q v_j, & \text{if } X = C \text{ and } j = l, \\ q^{-1} v_j, & \text{if } X = C \text{ and } j = l+1, \\ q v_j, & \text{if } X = D \text{ and } j \in \{l-1, l\}, \\ q^{-1} v_j, & \text{if } X = D \text{ and } j \in \{l+1, l+2\}, \\ v_j, & \text{otherwise}, \end{cases}$$

Via the coproduct, we get an action of $\mathbb{U}_q(\widetilde{\mathfrak{g}})$ on $V_K^{\otimes n}$.

Next we recall the definitions of specialized Birman–Murakami–Wenzl algebras and its action on $V_K^{\otimes n}$. Let r, x, q be three indeterminates over $\mathbb{Z}$.

Let R be the ring

$$R := \mathbb{Z}[r, r^{-1}, x, q, q^{-1}] / ((1 - x)(q - q^{-1}) + (r - r^{-1})).$$

For simplicity, we shall use the same letters r, r^{-1}, x, q, q^{-1} to denote their images in R respectively.

Definition 2.1. ([1,22]) The Birman–Murakami–Wenzl algebra $\mathfrak{B}_n(r, q)$ is a unital associative R-algebra with generators $T_i, E_i, 1 \leq i \leq n - 1$ and relations

1) $T_i - T_i^{-1} = (q - q_{\bullet}^{-1})(1 - E_i)$, for $1 \leq i \leq n - 1$,
2) $E_i^2 = xE_i$, for $1 \leq i \leq n - 1$,
3) $T_iT_{i+1}T_i = T_{i+1}T_iT_{i+1}$, for $1 \leq i \leq n - 2$,
4) $T_iT_j = T_jT_i$, for $|i - j| > 1$,
5) $E_iE_{i+1}E_i = E_i$, $E_{i+1}E_iE_{i+1} = E_{i+1}$, for $1 \leq i \leq n - 2$,
6) $T_iT_{i+1}E_i = E_{i+1}E_i$, $T_{i+1}T_iE_{i+1} = E_iE_{i+1}$, for $1 \leq i \leq n - 2$,
7) $E_iT_i = T_iE_i = r^{-1}E_i$, for $1 \leq i \leq n - 1$.
8) $E_iT_{i+1}E_i = rE_i$, $E_{i+1}T_iE_{i+1} = rE_{i+1}$, for $1 \leq i \leq n - 2$.

It is well known that $\mathfrak{B}_n(r, q)$ is a free R-module with rank $(2n - 1)!!$. If we specialize r to 1 and q to 1, then $\mathfrak{B}_n(r, q)$ will become the usual Brauer algebra with parameter x. We set $\mathscr{A} := \mathbb{Z}[q, q^{-1}]$ if $X \in \{C, D\}$; or $\mathscr{A} := \mathbb{Z}[q^{1/2}, q^{-1/2}]$ if $X = B$. We regard $\mathscr{A}$ as an R-algebra by sending r to q^{2l}, x to $1 + \sum_{i=-l}^{l-1} q^{2i+1}$ if $X = B$; or by sending r to $-q^{2l+1}$, x to $1 - \sum_{i=-l}^{l} q^{-2i}$ if $X = C$; or by sending r to q^{2l-1}, x to $1 + \sum_{i=1-l}^{l-1} q^{2i}$ if $X = D$. The resulting $\mathscr{A}$-algebra will be denoted by $\mathfrak{B}_{n,\mathscr{A}}$ and we set $\mathfrak{B}_{n,q} := \mathfrak{B}_{n,\mathscr{A}} \otimes_{\mathscr{A}} K$, see (1). We call it *the specialized Birman–Murakami–Wenzl algebra*, or specialized BMW algebra for short. Note that each T_i satisfies the following equation:

$$(T_i - q)(T_i + q^{-1})(T_i - r^{-1}) = 0. \tag{3}$$

If we specialize further q or $q^{1/2}$ to 1, then $\mathfrak{B}_{n,q}$ will become the specialized Brauer algebra used in [7,8]. By convention, we shall call the generators $E_1, \cdots, E_{n-1}$ the contraction generators of $\mathfrak{B}_{n,q}$.

There is an action of the algebra $\mathfrak{B}_{n,\mathscr{A}}$ on the n-tensor space $V_{\mathscr{A}}^{\otimes n}$ which we now recall. We set

$$(\rho_1, \cdots, \rho_N) := \begin{cases} (l - \frac{1}{2}, l - \frac{3}{2}, \cdots, \frac{1}{2}, 0, -\frac{1}{2}, \cdots, -l + \frac{3}{2}, -l + \frac{1}{2}), & \text{if } X = B; \\ (l, l - 1, \cdots, 1, -1, \cdots, -l + 1, -l), & \text{if } X = C; \\ (l - 1, l - 2, \cdots, 1, 0, 0, 1, \cdots, -l + 2, -l + 1), & \text{if } X = D. \end{cases}$$

If $X = C$ then we define $\epsilon_i := \text{sign}(\rho_i)$ for each $1 \leq i \leq 2l$. For any $i, j \in \{1, 2, \cdots, 2l\}$, we use $E_{i,j}$ to denote the corresponding basis of matrix unit for $\text{End}_K(V)$.

We define

$$\gamma' := \begin{cases} \displaystyle\sum_{1 \leq i,j \leq 2l} q^{\rho_j - \rho_i} \epsilon_i \epsilon_j E_{i,j'} \otimes E_{i',j}, & \text{if } X = C; \\ \displaystyle\sum_{1 \leq i,j \leq N} q^{\rho_j - \rho_i} E_{i,j'} \otimes E_{i',j}, & \text{if } X \in \{B, D\}. \end{cases}$$

We also define

$$\beta' := \sum_{1 \leq i \leq 2l} \left(q E_{i,i} \otimes E_{i,i} + q^{-1} E_{i,i'} \otimes E_{i',i} \right) + \sum_{\substack{1 \leq i,j \leq 2l \\ i \neq j, j'}} E_{i,j} \otimes E_{j,i}$$

$$+ (q - q^{-1}) \sum_{1 \leq i < j \leq 2l} \left(E_{i,i} \otimes E_{j,j} - q^{\rho_j - \rho_i} \epsilon_i \epsilon_j E_{i,j'} \otimes E_{i',j} \right),$$

if $X = C$; or

$$\beta' := \sum_{\substack{1 \leq i \leq N \\ i \neq i'}} \left(q E_{i,i} \otimes E_{i,i} + q^{-1} E_{i,i'} \otimes E_{i',i} \right) + \sum_{\substack{1 \leq i,j \leq N \\ i \neq j, j'}} E_{i,j} \otimes E_{j,i}$$

$$+ \delta_{N,2l+1} E_{l+1,l+1} \otimes E_{l+1,l+1}$$

$$+ (q - q^{-1}) \sum_{1 \leq i < j \leq N} \left(E_{i,i} \otimes E_{j,j} - q^{\rho_j - \rho_i} E_{i,j'} \otimes E_{i',j} \right),$$

if $X \in \{B, D\}$. Note that the operators β', γ' are related to each other by the equation

$$\beta' - (\beta')^{-1} = (q - q^{-1})(\text{id}_{V^{\otimes 2}} - \gamma').$$

For $i = 1, 2, \cdots, n - 1$, we set

$$\beta_i := \text{id}_{V^{\otimes i-1}} \otimes \beta' \otimes \text{id}_{V^{\otimes n-i-1}}, \quad \gamma_i := \text{id}_{V^{\otimes i-1}} \otimes \gamma' \otimes \text{id}_{V^{\otimes n-i-1}}.$$

By [15, §4] and [5, (10.2.5)], the map φ which sends each T_i to β_i and each E_i to γ_i for $i = 1, 2, \cdots, n-1$ can be naturally extended to a representation of $(\mathfrak{B}_{n,q})^{\text{op}}$ on $V^{\otimes n}$ such that all the statements in Theorem 1.1 hold. Note also that our β' is actually $(q^{-1}\beta_q)^{-1}$ in Hayashi's notations (see [15, §4]). If we specialize q or $q^{1/2}$ to 1, then this action of $\mathfrak{B}_{n,q}$ becomes the action studied in [7,8] of the specialized Brauer algebra $\mathfrak{B}_n(-N)$ or $\mathfrak{B}_n(N)$ on $V_{\mathbb{Z}}^{\otimes n}$.

3. Construction of maximal vectors

In this section we shall construct explicitly a complete set of maximal vectors in $V^{\otimes n}$.

By convention, a non-zero vector $v \in V^{\otimes n}$ is called a maximal vector if v is a maximal vector with respect to the action of $\mathbb{U}_q(\mathfrak{g})$, i.e., it satisfies that $E_i v = 0$ for any $1 \leq i \leq l$. For any two maximal vectors $v, w \in V^{\otimes n}$, we write $v \sim w$ if $v = wb$ or $w = vb$ for some $b \in \mathfrak{B}_{n,q}$. By a complete set of maximal vectors in $V^{\otimes n}$ we mean a complete set of representatives of maximal vectors under this equivalence relation "$\sim$".

For each sequence $\lambda = (\lambda_1, \lambda_2, \cdots, \lambda_l)$ of l integers, we define

$$\mathbf{n}(\lambda) := \begin{cases} (\lambda_1 - \lambda_2, \lambda_2 - \lambda_3, \cdots, \lambda_{l-1} - \lambda_l, \lambda_l), & \text{if } X = C; \\ (\lambda_1 - \lambda_2, \lambda_2 - \lambda_3, \cdots, \lambda_{l-1} - \lambda_l, 2\lambda_l), & \text{if } X = B; \\ (\lambda_1 - \lambda_2, \lambda_2 - \lambda_3, \cdots, \lambda_{l-1} - \lambda_l, \lambda_{l-1} + \lambda_l), & \text{if } X = D. \end{cases}$$

Recall that for any sequence $\lambda = (\lambda_1, \lambda_2, \cdots, \lambda_l)$ of l integers, a vector $v \in V^{\otimes n}$ is said to be a weight vector of weight λ, if $k_i v = q^{\mathbf{n}(\lambda)_i} v$ for each $1 \leq i \leq l$, where $(\mathbf{n}(\lambda)_1, \cdots, \mathbf{n}(\lambda)_l) := \mathbf{n}(\lambda)$. A weight $\lambda = (\lambda_1, \lambda_2, \cdots, \lambda_l)$ is said to be dominant if $\mathbf{n}(\lambda) \in (\mathbb{Z}^{\geq 0})^l$.

For each integer $k \geq 0$, we use $\Lambda^+(l, k)$ to denote the set of partitions of k with no more than l parts. Set

$$\Lambda^-(l, k) := \big\{ (\lambda_1, \cdots, \lambda_{l-1}, -\lambda_l) \mid (\lambda_1, \cdots, \lambda_l) \in \Lambda^+(l, k) \big\},$$
$$\Lambda_i^+(l, k) := \big\{ (\lambda_1, \cdots, \lambda_l) \in \Lambda^+(l, k) \mid \lambda_1, \cdots, \lambda_{l-i} \neq 0 \big\}, \text{for each } 0 \leq i \leq l.$$

We use $\pi^X(l, n)$ to denote the set of dominant weights in $V^{\otimes n}$. It is well known that (cf. [21])

$$\pi^X(l, n) = \begin{cases} \Big(\bigcup_{f \geq 0} \Lambda^+(l, n - 2f) \Big) \cup \Big(\bigcup_{f \geq 0} \Lambda_f^+(l, n - 2f - 1) \Big), & \text{if } X = B, \\ \bigcup_{f \geq 0} \Lambda^+(l, n - 2f), & \text{if } X = C, \\ \bigcup_{f \geq 0} \Lambda^{\pm}(l, n - 2f), & \text{if } X = D. \end{cases}$$

For each $\lambda \in \pi^X(l, n)$, we use $L(\lambda)$ to denote the irreducible $\mathbb{U}_q(\mathfrak{g})$-module with highest weight λ. We now explain the labelling of simple $\mathbb{U}_q(\widetilde{\mathfrak{g}})$-modules appeared in Theorem 1.1 in some details. The basic reference is [15]. If $X = C$, then $\widetilde{L}(\lambda) = L(\lambda)$; if $X = B$ and $\lambda_1' \leq l$, then $\widetilde{L}(\lambda) = L(\lambda)$ and θ acts on $\widetilde{L}(\lambda)$ as the scalar $(-1)^{|\lambda|}$; if $X = B$ and $\lambda_1' > l$, then $\widetilde{L}(\lambda) = L(\lambda^\dagger)$, where

$$\lambda^\dagger := \big(2l + 1 - \lambda_1', \lambda_2', \cdots, \lambda_{\lambda_1}' \big)',$$

and θ acts on $\widetilde{L}(\lambda)$ as the scalar $(-1)^{|\lambda^\dagger|+1}$; if $X = D$ and $\lambda_1' < l$, then $\widetilde{L}(\lambda) = L(\lambda)$ and θ acts on $\widetilde{L}(\lambda)$ as 1; if $X = D$ and $\lambda_1' > l$, then $\widetilde{L}(\lambda) = L(\lambda^\dagger)$, where

$$\lambda^\dagger := \left(2l - \lambda_1', \lambda_2', \cdots, \lambda_{\lambda_1}'\right)',$$

and θ acts on $\widetilde{L}(\lambda)$ as -1; if $X = D$ and $\lambda_1' = l$, then $\widetilde{L}(\lambda) \cong L(\lambda) \oplus L(\lambda)^\theta$, where $L(\lambda)^\theta \cong L(\lambda_1, \cdots, \lambda_{l-2}, \lambda_{l-1}, -\lambda_l)$.

Definition 3.1. We define

$$\alpha := \begin{cases} \displaystyle\sum_{1 \le k \le 2l} q^{-\rho_k} \epsilon_k v_k \otimes v_{k'}, & \text{if } X = C; \\ \displaystyle\sum_{1 \le k \le N} q^{-\rho_k} v_k \otimes v_{k'}, & \text{if } X \in \{B, D\}. \end{cases}$$

Lemma 3.1. $K\alpha$ *is a one dimensional trivial* $\mathbb{U}_q(\mathfrak{g})$*-submodule of* $V_{\mathbb{Q}(q)}^{\otimes 2}$.

Proof. By direct verification, we see that

$$\alpha = \begin{cases} q^{-1/2}(v_{l'} \otimes v_l)\gamma', & \text{if } X = B; \\ q^{-1}(v_{l'} \otimes v_l)\gamma', & \text{if } X = C; \\ (v_{l'} \otimes v_l)\gamma', & \text{if } X = D. \end{cases}$$

Since the right action of $\mathfrak{B}_{n,q}$ on $V^{\otimes n}$ commutes with the left action of $\mathbb{U}_q(\mathfrak{g})$, it follows that for any integer i with $1 \le i \le l$,

$$e_i \alpha = q^c e_i (v_{l'} \otimes v_l)\gamma' = q^c\left(e_i(v_{l'} \otimes v_l)\right)\gamma' = 0,$$

where $c \in \{-1/2, -1, 0\}$. Similarly, we have

$$f_i \alpha = q^c\left(f_i(v_{l'} \otimes v_l)\right)\gamma' = 0, \quad k_i \alpha = q^c\left(k_i(v_{l'} \otimes v_l)\right)\gamma' = \alpha.$$

If $X = B$, then it is clear that $\theta(\alpha) = \alpha$. If $X = D$, then

$$\theta(\alpha) = \left(\theta(v_{l'} \otimes v_l)\right)\gamma' = (v_l \otimes v_{l'})\gamma' = \alpha,$$

as required. This completes the proof of the lemma. $\quad\square$

Let $\mathcal{H}_q(\mathfrak{S}_n)$ be the Iwahori–Hecke algebra associated to the symmetric group $\mathfrak{S}_n$. By definition, it is a unital $\mathscr{A}$-algebra with generators $\widehat{T}_1, \cdots, \widehat{T}_{n-1}$ and relations:

$$(\widehat{T}_i - q)(\widehat{T}_i + q^{-1}) = 0, \quad \forall 1 \le i \le n-1;$$
$$\widehat{T}_i \widehat{T}_{i+1} \widehat{T}_i = \widehat{T}_{i+1} \widehat{T}_i \widehat{T}_{i+1}, \quad \forall 1 \le i < n-1,$$
$$\widehat{T}_i \widehat{T}_j = \widehat{T}_j \widehat{T}_i, \quad \forall 1 \le i < j-1 \le n-2.$$

For each $w \in \mathfrak{S}_n$ and each reduced expression $\sigma = s_{j_1} s_{j_2} \cdots s_{j_k}$ of w, we define

$$T_w = T_{j_1} T_{j_2} \cdots T_{j_k} \in \mathfrak{B}_{n,q}, \quad \widehat{T}_w = \widehat{T}_{j_1} \widehat{T}_{j_2} \cdots \widehat{T}_{j_k} \in \mathcal{H}_q(\mathfrak{S}_n),$$

Then the element T_w (resp., $\widehat{T}_w$) depends only on w and not on the choice of σ because of the braid relations. For each $1 \le i, j \le N$, we define

$$(v_i \otimes v_j)\widehat{\beta} := \begin{cases} q v_i \otimes v_i, & \text{if } i = j; \\ v_j \otimes v_i + (q - q^{-1}) v_i \otimes v_j, & \text{if } i < j; \\ v_j \otimes v_i, & \text{if } i > j. \end{cases}$$

For $i = 1, 2, \cdots, n - 1$, we set $\widehat{\beta}_i := \mathrm{id}_{V^{\otimes i-1}} \otimes \widehat{\beta} \otimes \mathrm{id}_{V^{\otimes n-i-1}}$. It is well known that the map $\widehat{\varphi}$ which sends each $\widehat{T}_i$ to $\widehat{\beta}_i$ for $i = 1, \cdots .n - 1$ can be naturally extended to a representation of $\mathcal{H}_q(\mathfrak{S}_n)$ on $V^{\otimes n}$.

If H is a standard Young subgroup of $\mathfrak{S}_n$, then we define

$$x_H := \sum_{w \in H} q^{\ell(w)} T_w, \quad y_H := \sum_{w \in H} (-q)^{-\ell(w)} T_w,$$

$$\widehat{x}_H := \sum_{w \in H} q^{\ell(w)} \widehat{T}_w, \quad \widehat{y}_H := \sum_{w \in H} (-q)^{-\ell(w)} \widehat{T}_w.$$

Then $\widehat{T}_i \widehat{x}_H = q \widehat{x}_H$, $\widehat{T}_i \widehat{y}_H = -q^{-1} \widehat{y}_H$, for any integer i satisfying $s_i \in H$. Let $0 \le f \le [\frac{n}{2}]$ be an integer and λ a partition of $n - 2f$. We set $\mathfrak{S}_\lambda := \mathfrak{S}_{\{2f+1,\cdots,2f+\lambda_1\}} \times \mathfrak{S}_{\{2f+\lambda_1+1,\cdots,2f+\lambda_1+\lambda_2\}} \times \cdots$, which is the Young subgroup of $\mathfrak{S}_{\{2f+1,\cdots,n\}}$ corresponding to λ, and we write

$$x_\lambda := x_{\mathfrak{S}_\lambda}, \quad y_\lambda := y_{\mathfrak{S}_\lambda}, \quad \widehat{x}_\lambda := \widehat{x}_{\mathfrak{S}_\lambda}, \quad \widehat{y}_\lambda := \widehat{y}_{\mathfrak{S}_\lambda}.$$

We denote by t^λ the standard λ-tableau in which the number $2f + 1, 2f + 2, \cdots, n$ appear in order along successive rows, t_λ the standard λ-tableau in which the number $2f + 1, 2f + 2, \cdots, n$ appear in order along successive columns. Let $w_\lambda \in \mathfrak{S}_{\{2f+1,\cdots,n\}}$ be such that $t^\lambda w_\lambda = t_\lambda$.

If λ is a partition satisfying $\ell(\lambda) \le N - l$, then we define

$$v_\lambda := \underbrace{v_1 \otimes \cdots \otimes v_1}_{\lambda_1 \text{ copies}} \otimes \underbrace{v_2 \otimes \cdots \otimes v_2}_{\lambda_2 \text{ copies}} \otimes \cdots \otimes \underbrace{v_{N-l} \otimes \cdots \otimes v_{N-l}}_{\lambda_{N-l} \text{ copies}}.$$

The important fact that we need is that $v_\lambda \mathfrak{B}_n^{(1)} = 0$ as long as v_{l+1} appears at most once in v_λ.

Definition 3.2. Let f be an integer with $0 \leq f \leq [n/2]$. Let λ be a partition of $n - 2f$ such that $\lambda \in \mathcal{P}_f(X)$. We set

$$\|\lambda\| := \begin{cases} n - 2f, & \text{if } X = C \text{ or } \lambda'_1 \leq l; \\ n - 2f - 2\lambda'_1 + 2l, & \text{if } X = D \text{ and } \lambda'_1 > l; \\ n - 2f - 2\lambda'_1 + 2l + 2, & \text{if } X = B \text{ and } \lambda'_1 > l. \end{cases}$$

If $X = B$ then $\lambda'_1 + \lambda'_2 \leq 2l + 1$. There are two possibilities:

Case b1. if $\lambda'_1 \leq l$ then we define

$$z_{f,\lambda} := \alpha^{\otimes f} \otimes v_\lambda T_{w_\lambda} y_{\lambda'}.$$

Case b2. if $\lambda'_1 > l$ then $\lambda'_2 \leq l$ and we set $a = \lambda'_1$, $c = \lambda_1$ and $\mu = (\mu_1, \cdots, \mu_c) := \lambda'$. We define

$$z_{f,\lambda} := \alpha^{\otimes f} \otimes \Big(\big(v_1 \otimes v_2 \otimes \cdots \otimes v_{2l+1-a} \otimes P_B(\lambda) \otimes v_{l+1} \big) \otimes \big(v_1 \otimes v_2 \otimes$$

$$\cdots \otimes v_{\mu_2} \big) \otimes \cdots \otimes \big(v_1 \otimes v_2 \otimes \cdots \otimes v_{\mu_c} \big) \Big) y_\mu,$$

where

$$P_B(\lambda) := \big(v_{(2l-a+2)'} \otimes v_{2l-a+2} - q^{-2(a-l-2)} v_{2l-a+2} \otimes v_{(2l-a+2)'} \big) \otimes$$

$$\big(v_{(2l-a+3)'} \otimes v_{2l-a+3} - q^{-2(a-l-3)} v_{2l-a+3} \otimes v_{(2l-a+3)'} \big) \otimes \cdots \otimes$$

$$\big(v_{(l-1)'} \otimes v_{l-1} - q^{-2} v_{l-1} \otimes v_{(l-1)'} \big) \otimes \big(v_{l'} \otimes v_l - v_l \otimes v_{l'} \big).$$

By convention, we understand that $P_B(\lambda) = \emptyset$ if $\lambda'_1 = l + 1$.

If $X = C$, then $\lambda'_1 \leq l$, we define

$$z_{f,\lambda} := \alpha^{\otimes f} \otimes v_\lambda T_{w_\lambda} y_{\lambda'}.$$

If $X = D$, then $\lambda'_1 + \lambda'_2 \leq 2l$. There are three possibilities:

Case d1. if $\lambda'_1 < l$ we define

$$z_{f,\lambda} := \alpha^{\otimes f} \otimes \big(v_\lambda T_{w_\lambda} y_{\lambda'} + \theta(v_\lambda T_{w_\lambda} y_{\lambda'}) \big).$$

Case d2. if $\lambda'_1 = l$ we define

$$z_{f,\lambda}^+ := \alpha^{\otimes f} \otimes v_\lambda T_{w_\lambda} y_{\lambda'}, \quad z_{f,\lambda}^- := \alpha^{\otimes f} \otimes \theta(v_\lambda T_{w_\lambda} y_{\lambda'}).$$

Case d3. if $\lambda'_1 > l$ then $\lambda'_2 < l$ and we set $a = \lambda'_1$, $c = \lambda_1$ and $\mu = (\mu_1, \cdots, \mu_c) := \lambda'$. We define

$$z_{f,\lambda} := \alpha^{\otimes f} \otimes \Big(\big(v_1 \otimes v_2 \otimes \cdots \otimes v_{2l-a} \otimes P_D(\lambda) \big) \otimes \big(v_1 \otimes v_2 \otimes$$

$$\cdots \otimes v_{\mu_2} \big) \otimes \cdots \otimes \big(v_1 \otimes v_2 \otimes \cdots \otimes v_{\mu_c} \big) \Big) y_\mu,$$

where

$$P_D(\lambda) := \big(v_{(2l-a+1)'} \otimes v_{2l-a+1} - q^{-2(a-l-1)} v_{2l-a+1} \otimes v_{(2l-a+1)'}\big) \otimes$$
$$\big(v_{(2l-a+2)'} \otimes v_{2l-a+2} - q^{-2(a-l-2)} v_{2l-a+2} \otimes v_{(2l-a+2)'}\big) \otimes \cdots \otimes$$
$$\big(v_{(l-1)'} \otimes v_{l-1} - q^{-2} v_{l-1} \otimes v_{(l-1)'}\big) \otimes \big(v_{l'} \otimes v_l - v_l \otimes v_{l'}\big).$$

The remaining part of this section is devoted to the proof that those $z_{f,\lambda}, z_{f,\lambda}^{\pm}$ give a complete set of maximal vectors in $V^{\otimes n}$.

Assume that $X = B$, $a = \lambda_1' > l$, $c = \lambda_1$ and $\mu = (\mu_1, \cdots, \mu_c) = \lambda'$. Then

$$z_{f,\lambda}\!\downarrow_{q^{1/2}=1} = 2^{a-l-1} \Big(\big(\alpha^{\otimes f} \otimes v_1 \otimes v_2 \otimes \cdots \otimes v_{2l+1-a} \otimes (v_{(2l-a+2)'} \otimes v_{2l-a+2})$$
$$\otimes (v_{(2l-a+3)'} \otimes v_{2l-a+3}) \otimes \cdots \otimes (v_{(l-1)'} \otimes v_{l-1}) \otimes (v_{l'} \otimes v_l)$$
$$\otimes v_{l+1}\big) \otimes (v_1 \otimes v_2 \otimes \cdots \otimes v_{\mu_2}) \otimes \cdots \otimes$$
$$(v_1 \otimes v_2 \otimes \cdots \otimes v_{\mu_c})) y_\mu \Big)\!\downarrow_{q^{1/2}=1}.$$

There is a similar equality for $z_{f,\lambda}\!\downarrow_{q=1}$ in the case where $X = D$ and $\lambda_1' > l$.

Lemma 3.2. *With the notations as in Definition 3.2, we have that $0 \neq z_{f,\lambda}, z_{f,\lambda}^{\pm} \in V_{\mathscr{A}}^{\otimes n}$. Furthermore,*

(1) if either $X = C$, or $X = B$ and $\lambda_1' \leq l$, or $X = D$ and $\lambda_1' < l$, then $z_{f,\lambda}$ is a weight vector of weight λ;

(2) if $X = D$ and $\lambda_1' = l$, then $z_{f,\lambda}^+$ is a weight vector of weight λ and $z_{f,\lambda}^-$ is a weight vector of weight $(\lambda_1, \cdots, \lambda_{l-1}, -\lambda_l)$;

(3) if $X \in \{B, D\}$ and $\lambda_1' > l$ then $z_{f,\lambda}$ is a weight vector of weight $\lambda^\dagger$;

Proof. It is clear that $z_{f,\lambda}, z_{f,\lambda}^{\pm} \in V_{\mathscr{A}}^{\otimes n}$. To prove that $z_{f,\lambda} \neq 0$, it suffices to show that $z_{f,\lambda}\!\downarrow_{q=1} \neq 0$. But the latter follows easily from the well known fact $v_\lambda T_{w_\lambda} y_{\lambda'}\!\downarrow_{q=1} \neq 0$, see, e.g., by [10, (2.8), Definition (3.3)]. The remaining claims all follow directly from the definitions of weights and of $z_{f,\lambda}, z_{f,\lambda}^{\pm}$. $\square$

We define $I(N, n) := \{\underline{i} = (i_1, \cdots, i_n) \mid 1 \leq i_j \leq N, \forall 1 \leq j \leq n\}$. For each $\underline{i} \in I(N, n)$, we write $v_{\underline{i}} := v_{i_1} \otimes \cdots \otimes v_{i_n} \in V^{\otimes n}$. The following very useful observation will be used many times in the rest of this section.

Lemma 3.3. *Let $\underline{i} = (i_1, \cdots, i_n) \in I(N, n)$. Let $1 \leq b < n$ be an integer such that $i_b \neq i_{b+1}'$. Then*

$$v_{\underline{i}} y_{(n)} = \begin{cases} -q v_{\underline{i} s_b} y_{(n)}, & \text{if } i_b > i_{b+1}, \\ -q^{-1} v_{\underline{i} s_b} y_{(n)}, & \text{if } i_b < i_{b+1}. \end{cases}$$

Proof. By definition,

$$y_{(n)} = \sum_{w \in \mathfrak{S}_n} (-q)^{-\ell(w)} T_w = -q^{-1}(T_b - q) \sum_{\substack{\sigma \in \mathfrak{S}_n \\ s_b \sigma > \sigma}} (-q)^{-\ell(\sigma)} T_\sigma.$$

Note that $(T_b + q^{-1})(T_b - q) = -(q - q^{-1})E_b T_b$. The lemma follows from direct verification. $\qquad\square$

Lemma 3.4. *1) If H is a standard Young subgroup of $\mathfrak{S}_n$, then $y_{(n)}$ has a left factor y_H.*

2) Let $\underline{i} = (i_1, \cdots, i_n) \in I(N, n)$. Let $1 \leq b < n$ be an integer such that $i_b = i_{b+1}$. Suppose that either $X \in \{C, D\}$ or $X = B$ and $i_b \neq l + 1$. Then $v_{\underline{i}} y_{(n)} = 0$.

Proof. 1) Let $\mathcal{D}$ be the set of minimal length distinguished right coset representatives of H in $\mathfrak{S}_n$. Then $y_{(n)} = y_H \times \left(\sum_{d \in \mathcal{D}} (-q)^{-\ell(d)} T_d \right)$, as required. This proves 1).

2) By 1), we see that $y_{(n)}$ has a left factor $-q^{-1}T_b + 1$. Therefore, the claim follows directly from the definition of the β' operator on $V^{\otimes n}$. $\qquad\square$

Lemma 3.5. *Let λ be a partition of n with $\ell(\lambda) \leq N - l$. Suppose that v_{l+1} appears at most once in v_λ. Then we have that*

$$v_\lambda T_{w_\lambda} y_{\lambda'} = v_\lambda w_\lambda y_{\lambda'} = v_\lambda w_\lambda \widehat{y}_{\lambda'}.$$

Moreover, $e_i v_\lambda T_{w_\lambda} y_{\lambda'} = 0$ for each $1 \leq i \leq l$.

Proof. By definition and our assumption on λ, it is clear that $v_\lambda T_w = v_\lambda \widehat{T}_w$ for any $w \in \mathfrak{S}_n$. In particular, $v_\lambda \widehat{x}_\lambda = \left(\sum_{w \in \mathfrak{S}_\lambda} q^{\ell(w)} \right) v_\lambda$.

Since $w_\lambda^{-1} = w_{\lambda'}$, we have that $\widehat{T}_{w_\lambda} - \widehat{T}_{w_{\lambda'}}^{-1} = \sum_{w < w_\lambda} a_w \widehat{T}_w$, where $a_w \in \mathscr{A}$ for each w. Applying [9, (4.1)], we get that

$$v_\lambda T_{w_\lambda} y_{\lambda'} = v_\lambda \widehat{T}_{w_\lambda} \widehat{y}_{\lambda'} = \left(\sum_{w \in \mathfrak{S}_\lambda} q^{\ell(w)} \right)^{-1} v_\lambda \widehat{x}_\lambda \widehat{T}_{w_\lambda} \widehat{y}_{\lambda'}$$

$$= \left(\sum_{w \in \mathfrak{S}_\lambda} q^{\ell(w)} \right)^{-1} v_\lambda \widehat{x}_\lambda \widehat{T}_{w_{\lambda'}}^{-1} \widehat{y}_{\lambda'} = v_\lambda \widehat{T}_{w_{\lambda'}}^{-1} \widehat{y}_{\lambda'}.$$

Note that

$$(v_i \otimes v_j)\widehat{\beta}^{-1} := \begin{cases} q^{-1} v_i \otimes v_i, & \text{if } i = j; \\ v_j \otimes v_i + (q^{-1} - q) v_i \otimes v_j, & \text{if } i > j; \\ v_j \otimes v_i, & \text{if } i < j. \end{cases}$$

It follows that (cf. [16, Lemma 3.9]) $v_\lambda \widehat{T}_{w_{\lambda'}}^{-1} = v_\lambda w_\lambda$. Hence $v_\lambda T_{w_\lambda} y_{\lambda'} = v_\lambda w_\lambda y_{\lambda'} = v_\lambda w_\lambda \widehat{y}_{\lambda'}$.

We write $\mu = (\mu_1, \cdots, \mu_c) = \lambda'$, where $c = \lambda_1$. Note that

$$v_\lambda w_\lambda = \underbrace{v_1 \otimes v_2 \otimes \cdots \otimes v_{\mu_1}}_{\mu_1 \text{ terms}} \otimes \underbrace{v_1 \otimes v_2 \otimes \cdots \otimes v_{\mu_2}}_{\mu_2 \text{ terms}} \otimes \cdots \otimes \underbrace{v_1 \otimes v_2 \otimes \cdots \otimes v_{\mu_c}}_{\mu_c \text{ terms}}.$$

For each integer i with $1 \leq i \leq l$, by direct calculation, we get that

$$\Delta^{(n-1)}(e_i) = e_i \otimes \underbrace{1 \otimes \cdots \otimes 1}_{n-1 \text{ copies}} + t_i \otimes e_i \otimes \underbrace{1 \otimes \cdots \otimes 1}_{n-2 \text{ copies}} + \cdots$$

$$+ t_i \otimes \cdots \otimes t_i \otimes e_i \otimes \underbrace{1}_{1 \text{ copies}} + t_i \otimes \cdots \otimes t_i \otimes e_i.$$

If $\ell(\lambda) \leq l$, then $l \geq \mu_1 \geq \cdots \geq \mu_c$. It follows that $e_i v_\lambda w_\lambda$ is a linear combination of some simple n-tensors of the form $v_{j^{(1)}} \otimes \cdots \otimes v_{j^{(c)}}$, where $v_{j^{(s)}} = v_{j_1^{(s)}} \otimes \cdots \otimes v_{j_{\mu_s}^{(s)}}$ for each $1 \leq s \leq l$, and such that $1 \leq v_{j_a^{(s)}} = v_{j_b^{(s)}} \leq l$ for some $1 \leq s \leq c$ and some integers $1 \leq a < b \leq \mu_s$. Each such simple n-tensor $v_{j^{(1)}} \otimes \cdots \otimes v_{j^{(c)}}$ is killed by $\widehat{y}_\mu$ (cf. [14]). This proves that $e_i v_\lambda T_{w_\lambda} \widehat{y}_{\lambda'} = 0$ in this case. If $\ell(\lambda) = l+1$ then $N = 2l+1$. By assumption, v_{l+1} appears exactly once in v_λ in this case. The remaining proof is similar. This completes the proof of the lemma. $\qquad\square$

Lemma 3.6. *With the notations as in Definition 3.2. We have that those $z_{f,\lambda}, z_{f,\lambda}^{\pm}$ form a complete set of maximal vectors in $V^{\otimes n}$.*

Proof. By Lemma 3.2 and the discussion of the labelling of simple $\mathbb{U}_q(\widetilde{\mathfrak{g}})$-modules appeared in Theorem 1.1 at the beginning of this section, we see that it suffices to prove that those $z_{f,\lambda}, z_{f,\lambda}^{\pm}$ are all maximal vectors in $V^{\otimes n}$.

If $X = C$, or $X = B$ and $\lambda_1' \leq l$, or $X = D$ and $\lambda_1' < l$, then it follows directly from Lemma 3.5 and Lemma 3.1 that $z_{f,\lambda}$ is a maximal vector in $V^{\otimes n}$. If $X = D$ and $\lambda_1' = l$, it is also clear that (by Lemma 3.5 and Lemma 3.1 and the commutator relations between e_i and θ) that $z_{f,\lambda}^{\pm}$ are maximal vectors in $V^{\otimes n}$. It remains to consider the case where either $X = B$ and $\lambda_1' > l$, or $X = D$ and $\lambda_1' > l$. As before, we write $a = \lambda_1'$, $c = \lambda_1$ and $\mu = (\mu_1, \cdots, \mu_c) = \lambda'$.

Case 1. $X = B$ and $a > l$. Let b be an integer with $1 \leq b \leq l$. By Lemma 3.4, the proof of Lemma 3.5 and the definition of $z_{f,\lambda}$, it is easy to see that to prove $e_b z_{f,\lambda} = 0$, it suffices to show that

$$e_b \left(\left(v_1 \otimes v_2 \otimes \cdots \otimes v_{2l+1-a} \otimes P_B(\lambda) \otimes v_{l+1} \right) y_{(a)} \right) = 0. \tag{4}$$

If $a = l+1$ or $a > l+1$ and $1 \le b \le 2l - a$, then by the same argument used in the proof of Lemma 3.5 and the fact that $y_{(a)}$ has a left factor $y_{(2l+1-a)}$, we see easily that the left-hand side of (4) is equal to 0. Thus it suffices to consider the case when $a > l+1$ and $2l + 1 - a \le b \le l$. If $b = 2l + 1 - a$ and $a > l+1$, then

$$\left(e_b \left(v_1 \otimes v_2 \otimes \cdots \otimes v_{2l+1-a} \otimes P_B(\lambda) \otimes v_{l+1} \right) \right) y_{(a)}$$

$$= \left(e_b \left(v_1 \otimes v_2 \otimes \cdots \otimes v_{2l+1-a} \otimes (v_{(2l-a+2)'} \otimes v_{2l-a+2} \right. \right.$$
$$\left. \left. - q^{-2(a-l-2)} v_{2l-a+2} \otimes v_{(2l-a+2)'}) \right) \otimes F_1 \otimes v_{l+1} \right) y_{(a)}$$

$$= q^{1/2} \left(\left(v_1 \otimes v_2 \otimes \cdots \otimes v_{2l+1-a} \right) \otimes \left(q^{1/2} v_{(2l-a+2)'} \otimes v_{2l-a+1} \right. \right.$$
$$\left. \left. - q^{-2(a-l-2)} v_{2l-a+1} \otimes v_{(2l-a+2)'} \right) \otimes F_1 \otimes v_{l+1} \right) y_{(a)}$$

$$= q^{1/2} \left(\left(v_1 \otimes v_2 \otimes \cdots \otimes v_{2l+1-a} \right) \otimes \left(-q^{3/2} v_{2l-a+1} \otimes v_{(2l-a+2)'} \right. \right.$$
$$\left. \left. - q^{-2(a-l-2)} v_{2l-a+1} \otimes v_{(2l-a+2)'} \right) \otimes F_1 \otimes v_{l+1} \right) y_{(a)}$$

$$= 0,$$

where F_1 is a linear combination of some simple $(2a - 2l - 4)$-tensors $v_{p_1} \otimes \cdots \otimes v_{p_{2a-2l-4}}$ which satisfies $2l - a + 3 \le p_t \le (2l - a + 3)'$ for each t, and we have used Lemmas 3.3 and 3.4 in the third and the fourth equalities respectively.

Therefore, it is enough to consider the case when $a > l+1$ and $2l+2-a \le b \le l$. Equivalently, it suffices to show that

$$e_b \left(\left(P_B(\lambda) \otimes v_{l+1} \right) y_{(2a-2l-1)} \right) = 0, \tag{5}$$

for each $2l + 2 - a \le b \le l$.

We use induction on $a - l$. If $a - l = 2$, then $b = l$ and the left-hand side of (5) becomes

$$e_l \left(\left((v_{l'} \otimes v_l - v_l \otimes v_{l'}) \otimes v_{l+1} \right) y_{(3)} \right)$$

$$= e_l \left(\left(v_{l'} \otimes v_l \otimes v_{l+1} - v_{l+1} \otimes v_l \otimes v_{l'} \right) y_{(3)} \right)$$

$$= \left(-v_{l+1} \otimes v_l \otimes v_{l+1} + q^{1/2} v_{l'} \otimes v_l \otimes v_l - q^{1/2} v_l \otimes v_l \otimes v_{l'} \right.$$
$$\left. + v_{l+1} \otimes v_l \otimes v_{l+1} \right) y_{(3)}$$

$$= 0,$$

where we have used again Lemmas 3.3 and 3.4 in the first and the third equalities respectively.

Now assume that

$$e_b\left(\left(\bigotimes_{t=0}^{a-l-2}(v_{(l-t)'}\otimes v_{l-t}-q^{-2t}v_{l-t}\otimes v_{(l-t)'})\otimes v_{l+1}\right)y_{(2a-2l-1)}\right)=0,$$

for each $2l+2-a\le b\le l$, where the tensor product are read from left to right with decreasing values of t. We are going to prove that

$$e_b\left(\left(\bigotimes_{t=0}^{a-l-1}(v_{(l-t)'}\otimes v_{l-t}-q^{-2t}v_{l-t}\otimes v_{(l-t)'})\otimes v_{l+1}\right)y_{(2a-2l+1)}\right)=0,$$

for each $2l-a+1\le b\le l$, where the tensor product are read from left to right with decreasing values of t. Clearly, it suffices to prove the above equality for $b=2l-a+1$ (as for other b the equality already follows from induction assumption).

Note that

$$e_{2l-a+1}\left(\left(v_{(2l-a+1)'}\otimes v_{2l-a+1}-q^{-2(a-l-1)}v_{2l-a+1}\otimes v_{(2l-a+1)'}\right)\right.$$

$$\left.\otimes\left(v_{(2l-a+2)'}\otimes v_{2l-a+2}-q^{-2(a-l-2)}v_{2l-a+2}\otimes v_{(2l-a+2)'}\right)\right)y_{(4)}$$

$$=e_{2l-a+1}\left(v_{(2l-a+1)'}\otimes v_{2l-a+1}\otimes\left(v_{(2l-a+2)'}\otimes v_{2l-a+2}\right.\right.$$

$$\left.\left.-q^{-2(a-l-2)}v_{2l-a+2}\otimes v_{(2l-a+2)'}\right)\right)y_{(4)}-q^{-2(a-l-1)}e_{2l-a+1}\left(v_{2l-a+1}\right.$$

$$\left.\otimes v_{(2l-a+1)'}\otimes\left(v_{(2l-a+2)'}\otimes v_{2l-a+2}-q^{-2(a-l-2)}v_{2l-a+2}\otimes v_{(2l-a+2)'}\right)\right)y_{(4)}$$

$$=e_{2l-a+1}\left(v_{(2l-a+1)'}\otimes v_{2l-a+1}\otimes\left(v_{(2l-a+2)'}\otimes v_{2l-a+2}\right.\right.$$

$$\left.\left.-q^{-2(a-l-2)}v_{2l-a+2}\otimes v_{(2l-a+2)'}\right)\right)y_{(4)}+q^{-2(a-l)+3}e_{2l-a+1}\left(v_{(2l-a+2)'}\right.$$

$$\left.\otimes v_{2l-a+1}\otimes v_{2l-a+2}\otimes v_{(2l-a+1)'}\right)y_{(4)}-q^{-4(a-l)+7}e_{2l-a+1}\left(v_{2l-a+2}\right.$$

$$\left.\otimes v_{2l-a+1}\otimes v_{(2l-a+2)'}\otimes v_{(2l-a+1)'}\right)y_{(4)}$$

$$=q^{-2(a-l-2)}\left(v_{(2l-a+2)'}\otimes v_{2l-a+1}\otimes v_{2l-a+2}\otimes v_{(2l-a+2)'}\right)y_{(4)}$$

$$+q^{-2(a-l)+3}q(-1)\left(v_{(2l-a+2)'}\otimes v_{2l-a+1}\otimes v_{2l-a+2}\otimes v_{(2l-a+2)'}\right)y_{(4)}$$

$$=0,$$

where we have used again Lemmas 3.3 and 3.4 in the last three equalities. Since $e_{2l-a+1}v_j=0$ for any $2l-a+3\le j\le(2l-a+3)'$, it follows from the above equality that

$$e_{2l-a+1}\left(\left(\bigotimes_{t=0}^{a-l-1}(v_{(l-t)'}\otimes v_{l-t}-q^{-2t}v_{l-t}\otimes v_{(l-t)'})\otimes v_{l+1}\right)y_{(2a-2l+1)}\right)=0,$$

as required. This completes the proof in Case 1.

Case 2. $X = D$ and $a > l$. The proof is actually similar to that in Case 1. As in Case 1, we see easily that it suffices to show that

$$e_b\Big(\big(v_1 \otimes v_2 \otimes \cdots \otimes v_{2l-a} \otimes P_D(\lambda)\big)y_{(a)}\Big) = 0, \tag{6}$$

for each $2l - a \le b \le l$. If $b = 2l - a$, then

$$\Big(e_b\big(v_1 \otimes v_2 \otimes \cdots \otimes v_{2l-a} \otimes P_D(\lambda)\big)\Big)y_{(a)}$$

$$= \Big(e_b\big(v_1 \otimes v_2 \otimes \cdots \otimes v_{2l-a} \otimes (v_{(2l-a+1)'} \otimes v_{2l-a+1}$$
$$- q^{-2(a-l-1)}v_{2l-a+1} \otimes v_{(2l-a+1)'})\big) \otimes F_2\Big)y_{(a)}$$

$$= q\Big(\big(v_1 \otimes v_2 \otimes \cdots \otimes v_{2l-a}\big) \otimes \big(qv_{(2l-a+1)'} \otimes v_{2l-a}$$
$$- q^{-2(a-l-1)}v_{2l-a} \otimes v_{(2l-a+1)'}\big) \otimes F_2\Big)y_{(a)}$$

$$= q\Big(\big(v_1 \otimes v_2 \otimes \cdots \otimes v_{2l-a}\big) \otimes \big(-q^2 v_{2l-a} \otimes v_{(2l-a+1)'}$$
$$- q^{-2(a-l-1)}v_{2l-a} \otimes v_{(2l-a+1)'}\big) \otimes F_2\Big)y_{(a)}$$

$$= 0,$$

where F_2 is a linear combination of some simple $(2a - 2l - 2)$-tensors $v_{p_1} \otimes \cdots \otimes v_{p_{2a-2l-2}}$ which satisfies $2l - a + 2 \le p_t \le (2l - a + 2)'$ for each t, and we have used Lemmas 3.3 and 3.4 in the third and the fourth equalities respectively. Therefore, it is enough to consider the case when $2l + 1 - a \le b \le l$. If $a - l = 1$, then $b = l$ and the left-hand side of (6) becomes

$$e_l\Big(v_1 \otimes v_2 \otimes \cdots \otimes v_{l-1} \otimes \big((v_{l'} \otimes v_l - v_l \otimes v_{l'})\big)y_{(l+1)}\Big)$$

$$= \Big(v_1 \otimes v_2 \otimes \cdots \otimes v_{l-1} \otimes \big(v_{l-1} \otimes v_l - qv_l \otimes v_{l-1}\big)\Big)y_{(l+1)}$$

$$= \Big(v_1 \otimes v_2 \otimes \cdots \otimes v_{l-1} \otimes \big(v_{l-1} \otimes v_l + q^2 v_{l-1} \otimes v_l\big)\Big)y_{(l+1)}$$

$$= 0,$$

where we have used again Lemmas 3.3 and 3.4 in the second and the third equalities respectively.

Therefore, it is enough to consider the case where $a - l > 1$ and $2l + 1 - a \le b \le l$. Equivalently, it suffices to show that

$$e_b\Big(\big(P_D(\lambda)\big)y_{(2a-2l)}\Big) = 0, \tag{7}$$

for each $2l + 1 - a \le b \le l$.

We use induction on $a - l$. We assume that

$$e_b\left(\left(\bigotimes_{t=0}^{a-l-1} (v_{(l-t)'} \otimes v_{l-t} - q^{-2t}v_{l-t} \otimes v_{(l-t)'})\right)y_{(2a-2l)}\right) = 0,$$

for each $2l + 1 - a \le b \le l$, where the tensor product are read from left to right with decreasing values of t. We are going to prove that

$$e_b\left(\left(\bigotimes_{t=0}^{a-l} (v_{(l-t)'} \otimes v_{l-t} - q^{-2t}v_{l-t} \otimes v_{(l-t)'})\right)y_{(2a-2l+2)}\right) = 0,$$

for each $2l - a \le b \le l$, where the tensor product are read from left to right with decreasing values of t. Clearly, it suffices to prove the above equality for $b = 2l - a$ (as for other b the equality already follows from induction assumption).

Note that

$$e_{2l-a}\left(\left(v_{(2l-a)'} \otimes v_{2l-a} - q^{-2(a-l)}v_{2l-a} \otimes v_{(2l-a)'}\right)\right.$$

$$\left. \otimes \left(v_{(2l-a+1)'} \otimes v_{2l-a+1} - q^{-2(a-l-1)}v_{2l-a+1} \otimes v_{(2l-a+1)'}\right)\right)y_{(4)}$$

$$= e_{2l-a}\left(v_{(2l-a)'} \otimes v_{2l-a} \otimes \left(v_{(2l-a+1)'} \otimes v_{2l-a+1}\right.\right.$$

$$\left.\left. - q^{-2(a-l-1)}v_{2l-a+1} \otimes v_{(2l-a+1)'}\right)\right)y_{(4)} - q^{-2(a-l)}e_{2l-a}\left(v_{2l-a}\right.$$

$$\left. \otimes v_{(2l-a)'} \otimes \left(v_{(2l-a+1)'} \otimes v_{2l-a+1} - q^{-2(a-l-1)}v_{2l-a+1} \otimes v_{(2l-a+1)'}\right)\right)y_{(4)}$$

$$= e_{2l-a}\left(v_{(2l-a)'} \otimes v_{2l-a} \otimes \left(v_{(2l-a+1)'} \otimes v_{2l-a+1}\right.\right.$$

$$\left.\left. - q^{-2(a-l-1)}v_{2l-a+1} \otimes v_{(2l-a+1)'}\right)\right)y_{(4)} + q^{-2(a-l)+1}e_{2l-a}\left(v_{(2l-a+1)'}\right.$$

$$\left. \otimes v_{2l-a} \otimes v_{2l-a+1} \otimes v_{(2l-a)'}\right)y_{(4)} - q^{-4(a-l)+3}e_{2l-a}\left(v_{2l-a+1}\right.$$

$$\left. \otimes v_{2l-a} \otimes v_{(2l-a+1)'} \otimes v_{(2l-a)'}\right)y_{(4)}$$

$$= q^{-2(a-l-1)}\left(v_{(2l-a+1)'} \otimes v_{2l-a} \otimes v_{2l-a+1} \otimes v_{(2l-a+1)'}\right)y_{(4)}$$

$$+ q^{-2(a-l)+1}q(-1)\left(v_{(2l-a+1)'} \otimes v_{2l-a} \otimes v_{2l-a+1} \otimes v_{(2l-a+1)'}\right)y_{(4)}$$

$$= 0,$$

where we have used again Lemmas 3.3 and 3.4 in the last three equalities. Since $e_{2l-a}v_j = 0$ for any $2l - a + 2 \le j \le (2l - a + 2)'$, it follows from the above equality that

$$e_{2l-a}\left(\left(\bigotimes_{t=0}^{a-l} (v_{(l-t)'} \otimes v_{l-t} - q^{-2t}v_{l-t} \otimes v_{(l-t)'}) \otimes v_{l+1}\right)y_{(2a-2l+2)}\right) = 0,$$

as required. This completes the proof in Case 2 and hence the proof of the lemma. □

4. Proof of Theorem 1.2, Theorem 1.3 and Theorem 1.4

The purpose of this section is to give a proof of the Theorem 1.2, Theorem 1.3 and Theorem 1.4.

Let $\underline{i} = (i_1, \cdots, i_n) \in I(N, n)$. An ordered pair (s, t) with $1 \leq s < t \leq n$ is called an X-pair in $\underline{i}$ if $i_s = (i_t)'$. Two ordered pairs (s, t) and (u, v) are called disjoint if $\{s, t\} \cap \{u, v\} = \emptyset$. We define the X-length $\ell_X(v_{\underline{i}}) = \ell_X(\underline{i})$ to be the maximal number of disjoint X-pairs (s, t) in $\underline{i}$. The following simple observation is useful.

Lemma 4.1. *For any* $\underline{i} \in I(N, n)$ *and* $b \in \mathfrak{B}_{n,q}$, $v_{\underline{i}}b$ *is a linear combination of some simple n-tensors $v_{\underline{j}}$ such that $\ell_X(\underline{j}) = \ell_X(\underline{i})$.*

Proof. This follows directly from the definition of the action of $\mathfrak{B}_{n,q}$ on $V^{\otimes n}$. □

For each integer $0 \leq f \leq [n/2]$, let $\mathfrak{B}_n^{(f)}$ be the two-sided ideal of $\mathfrak{B}_{n,q}$ which is generated by $E_1 E_3 \cdots E_{2f-1}$. Thus we have a chain of ideals

$$\mathfrak{B}_{n,q} = \mathfrak{B}_n^{(0)} \supseteq \mathfrak{B}_n^{(1)} \supseteq \cdots \supseteq \mathfrak{B}_n^{([\frac{n}{2}])} \supseteq 0,$$

which gives rise to a filtration of the n-tensor space

$$V^{\otimes n} = V^{\otimes n}\mathfrak{B}_n^{(0)} \supseteq V^{\otimes n}\mathfrak{B}_n^{(1)} \supseteq \cdots \supseteq V^{\otimes n}\mathfrak{B}_n^{([\frac{n}{2}])} \supseteq 0.$$

For each integer $0 \leq f \leq [n/2]$ and $\lambda \in \mathcal{P}_f(X)$, we set

$$D_{g,\lambda} := \widetilde{D}_{f,\lambda}, \tag{8}$$

whenever $\|\lambda\| = n - 2g$.

Proposition 4.1. *For each integer $0 \leq f \leq [n/2]$, there is an irreducible $\mathbb{U}_q(\widetilde{\mathfrak{g}})$-$\mathfrak{B}_{n,q}$-bimodule decomposition*

$$V^{\otimes n}\mathfrak{B}_n^{(f)} = \bigoplus_{f \leq g \leq [n/2]} \bigoplus_{\substack{\lambda \in \mathcal{P}(X) \\ \|\lambda\| = n - 2g}} \widetilde{L}(\lambda) \otimes D_{g,\lambda}.$$

Proof. By Theorem 1.1, we have a $\mathbb{U}_q(\widetilde{\mathfrak{g}})$-$\mathfrak{B}_{n,q}$-bimodule decomposition

$$V^{\otimes n} = \bigoplus_{g=0}^{[n/2]} \bigoplus_{\lambda \in \mathcal{P}_g(X)} \widetilde{L}(\lambda) \otimes \widetilde{D}_{g,\lambda}.$$

In particular, this implies that any nonzero maximal vector in $V^{\otimes n}$ generates a copy of a simple right $\mathfrak{B}_{n,q}$-module.

Let g be an integer with $0 \le g \le [n/2]$ and $\lambda \in \mathcal{P}_g(X)$. There are three cases:

Case 1. $X \in \{B, C\}$ and $\lambda_1' \le l$ or $X = D$ and $\lambda_1' < l$. In this case, we have that $|\lambda| = \| \lambda \|$ and $z_{g,\lambda} \in V^{\otimes n}\mathfrak{B}_n^{(g)}$. By Lemma 3.6, $z_{g,\lambda}$ is a nonzero maximal vector of weight λ in $V^{\otimes n}$ and $\mathbb{U}_q(\widetilde{\mathfrak{g}})z_{g,\lambda} \cong \widetilde{L}(\lambda)$. Hence $z_{g,\lambda}\mathfrak{B}_{n,q} \cong \widetilde{D}_{g,\lambda}$. Since $\widetilde{L}(\lambda) \otimes \widetilde{D}_{g,\lambda}$ is an irreducible $\mathbb{U}_q(\widetilde{\mathfrak{g}})$-$\mathfrak{B}_{n,q}$-bimodule which appears with multiplicity one in $V^{\otimes n}$ and $\mathbb{U}_q(\widetilde{\mathfrak{g}})z_{g,\lambda}\mathfrak{B}_{n,q}$ is a $\mathbb{U}_q(\widetilde{\mathfrak{g}})$-$\mathfrak{B}_{n,q}$-subbimodule of $V^{\otimes n}$, it follows that $\mathbb{U}_q(\widetilde{\mathfrak{g}})z_{g,\lambda}\mathfrak{B}_{n,q} \cong \widetilde{L}(\lambda) \otimes \widetilde{D}_{g,\lambda} = \widetilde{L}(\lambda) \otimes D_{g,\lambda}$.

Case 2. $X \in \{B, D\}$ and $\lambda_1' > l$. In this case, $|\lambda| \ge \|\lambda\|$. By Lemma 3.6, $z_{g,\lambda}$ is a nonzero maximal vector of weight $\lambda^\dagger$ in $V^{\otimes n}$ and $\mathbb{U}_q(\widetilde{\mathfrak{g}})z_{g,\lambda} \cong \widetilde{L}(\lambda)$. By a similar argument as used in Case 1, we still have that

$$\mathbb{U}_q(\widetilde{\mathfrak{g}})z_{g,\lambda}\mathfrak{B}_{n,q} \cong \widetilde{L}(\lambda) \otimes D_{g,\lambda}.$$

Furthermore, by the definition of $z_{g,\lambda}$, we can find integers $1 \le i_1 < i_2 < \cdots < i_g$ such that $i_{s+1} - i_s > 1$ for each $1 \le s < g$, and $z_{g,\lambda}E_{i_1}E_{i_2}\cdots E_{i_g} \ne 0$. As $z_{g,\lambda}\mathfrak{B}_{n,q}$ is a simple right $\mathfrak{B}_{n,q}$-module, it follows that $z_{g,\lambda}\mathfrak{B}_{n,q} = z_{g,\lambda}E_{i_1}E_{i_2}\cdots E_{i_g}\mathfrak{B}_{n,q} \subseteq V^{\otimes n}\mathfrak{B}_n^{(g)}$. As a result, we get that $z_{g,\lambda} \in V^{\otimes n}\mathfrak{B}_n^{(g)}$

Case 3. $X = D$ and $\lambda_1' = l$. In this case, we also have that $|\lambda| = \|\lambda\|$ and $z_{g,\lambda}^{\pm} \in V^{\otimes n}\mathfrak{B}_n^{(g)}$. By Lemma 3.6, $z_{g,\lambda}^+, z_{g,\lambda}^-$ are nonzero maximal vectors of weight λ and $(\lambda_1, \cdots, \lambda_{l-1}, -\lambda_l)$ in $V^{\otimes n}$ respectively, and $\mathbb{U}_q(\mathfrak{g})z_{g,\lambda}^+ \oplus \mathbb{U}_q(\mathfrak{g})z_{g,\lambda}^- \cong \widetilde{L}(\lambda)$. Note that $\theta z_{g,\lambda}^+\mathfrak{B}_{n,q} = z_{g,\lambda}^-\mathfrak{B}_{n,q} \cong \widetilde{D}_{g,\lambda}$. By a similar argument as used in Case 1, we still have that

$$\left(\mathbb{U}_q(\mathfrak{g})z_{g,\lambda}^+ \oplus \mathbb{U}_q(\mathfrak{g})z_{g,\lambda}^-\right)\mathfrak{B}_{n,q} \cong \widetilde{L}(\lambda) \otimes \widetilde{D}_{g,\lambda} = \widetilde{L}(\lambda) \otimes D_{g,\lambda}.$$

From the above discussion on three cases, we deduce that

$$\bigoplus_{g=f}^{[n/2]} \bigoplus_{\substack{\lambda \in \mathcal{P}(X) \\ \|\lambda\|=n-2g}} \widetilde{L}(\lambda) \otimes D_{g,\lambda} \subseteq V^{\otimes n}\mathfrak{B}_n^{(f)}.$$

We claim that the above inclusion "$\subseteq$" must be an equality. Suppose this is not the case. Since the above inclusion "$\subseteq$" is an equality when $f = 0$, it follows that there must exist some integer $0 \le g < f$ and some partition $\mu \in \mathcal{P}(X)$ with $\| \mu \| = n - 2g$ such that $\widetilde{L}(\mu) \otimes D_{g,\mu} \subseteq V^{\otimes n}\mathfrak{B}_n^{(f)}$. By the above analysis on three cases, we see that this means $z_{(n-|\mu|)/2,\mu} \in V^{\otimes n}\mathfrak{B}_n^{(f)}$ or $z_{(n-|\mu|)/2,\mu}^{\pm} \in V^{\otimes n}\mathfrak{B}_n^{(f)}$. However, by definition, $z_{(n-|\mu|)/2,\mu}$

(resp., $z^{\pm}_{(n-|\mu|)/2,\mu}$) is a linear combination of some simple n-tensors which has X-length g, while every element in $V^{\otimes n}\mathfrak{B}_n^{(f)}$ is a linear combination of some simple n-tensors which has X-length at least $f > g$. We get a contradiction. This completes the proof of the proposition. $\qquad\square$

Corollary 4.1. *Let f be an integer with $0 \leq f \leq [n/2]$. Then there is a $\mathbb{U}_q(\widetilde{\mathfrak{g}})$-$\mathfrak{B}_{n,q}$-bimodule isomorphism:*

$$V^{\otimes n}\mathfrak{B}_n^{(f)}/V^{\otimes n}\mathfrak{B}_n^{(f+1)} \cong \bigoplus_{\substack{\lambda\in\mathcal{P}(X) \\ \|\lambda\|=n-2f}} \widetilde{L}(\lambda) \otimes D_{f,\lambda}.$$

Let f be an integer with $0 \leq f \leq [n/2]$. Recall the definition of $\mathcal{H}\mathcal{T}_{f,q}^{\otimes n}$ in the introduction.

Proof of Theorem 1.2: It is easy to see that $\mathcal{H}\mathcal{T}_{f,q}^{\otimes n}$ is a $\mathbb{U}_q(\widetilde{\mathfrak{g}})$-$\mathfrak{B}_{n,q}$ sub-bimodule of $V^{\otimes n}$. We identify each $\widetilde{L}(\lambda) \otimes \widetilde{D}_{f,\lambda}$ as a subspace of $V^{\otimes n}$ using the isomorphism given in Theorem 1.1. By the proof of Proposition 4.1, we get that

$$\bigoplus_{\substack{\lambda\in\mathcal{P}(X) \\ \|\lambda\|=n-2f}} \widetilde{L}(\lambda) \otimes D_{f,\lambda} \subseteq \mathcal{H}\mathcal{T}_{f,q}^{\otimes n}.$$

We claim that the above inclusion "$\subseteq$" must be an equality. Suppose this is not the case. Then there must exist some integer $0 \leq g \neq f \leq [n/2]$ and some partition μ satisfying $\| \mu \| = n - 2g$ such that

$$\widetilde{L}(\mu) \otimes D_{g,\mu} \subseteq \mathcal{H}\mathcal{T}_{f,q}^{\otimes n} \subseteq V^{\otimes n}\mathfrak{B}_n^{(f)}.$$

It follows that $z_{(n-|\mu|)/2,\mu} \in V^{\otimes n}\mathfrak{B}_n^{(f)}$ and hence $g > f$. If $X \in \{B,C\}$ and $\mu_1' \leq l$, then $|\mu| = \|\mu\| = n - 2g$ and we have that

$$z_{g,\mu}E_1 E_3 \cdots E_{2g-1} = (\alpha^{\otimes g})E_1 E_3 \cdots E_{2g-1} \otimes v_\mu T_{w_\mu} y_{\mu'} = x^g z_{g,\mu} \neq 0.$$

If $X = D$ and $\mu_1' < l$, or $X \in \{B,D\}$ and $\mu_1' > l$, using the definition of $z_{(n-|\mu|)/2,\mu}$ given in Definition 3.2, we can also find integers $1 \leq i_1, i_2, \cdots, i_g < n$ such that $i_{s+1} - i_s > 1$ for each $1 \leq s < g$ and $z_{(n-|\mu|)/2,\mu}E_{i_1}E_{i_2} \cdots E_{i_g} \neq 0$. The same statement applies to $z^{\pm}_{g,\mu}$ in the case where $X = D$ and $\lambda_1' = l$. Therefore, we get a contradiction to fact that $\mathcal{H}\mathcal{T}_{f,q}^{\otimes n}E_{i_1} \cdots E_{i_g} = 0$. This proves our claim.

Combining this claim with Corollary 4.1, we get that

$$\mathcal{H}\mathcal{T}_{f,q}^{\otimes n} = \bigoplus_{\substack{\lambda\in\mathcal{P}(X) \\ \|\lambda\|=n-2f}} \widetilde{L}(\lambda) \otimes \widetilde{D}_{(n-|\lambda|)/2,\lambda} \cong V^{\otimes n}\mathfrak{B}_n^{(f)}/V^{\otimes n}\mathfrak{B}_n^{(f+1)}.$$

As a result, we have a $\mathbb{U}_q(\widetilde{\mathfrak{g}})$-$\mathfrak{B}_{n,q}$-bimodule decomposition:

$$V^{\otimes n} = \bigoplus_{f=0}^{[n/2]} \mathcal{HT}_{f,q}^{\otimes n}.$$

This completes the proof of Theorem 1.2.

Proof of Theorem 1.3: By [11, (3.2)], we know that there is an algebra isomorphism $\mathcal{H}_q(\mathfrak{S}_n) \cong \mathfrak{B}_{n,q}/\mathfrak{B}_n^{(1)}$. Furthermore, the right action of $\mathfrak{B}_{n,q}$ on $\mathcal{HT}_{0,q}^{\otimes n}$ factors through to a right action of $\mathfrak{B}_{n,q}/\mathfrak{B}_n^{(1)}$ and hence of $\mathcal{H}_q(\mathfrak{S}_n)$ on $\mathcal{HT}_{0,q}^{\otimes n}$. In particular, we see that the natural left action of $\mathbb{U}_q(\widetilde{\mathfrak{g}})$ on $\mathcal{HT}_{0,q}^{\otimes n}$ commutes with the right action of $\mathcal{H}_q(\mathfrak{S}_n)$. The statement 2) follows as a special case of Corollary 4.1 and Theorem 1.2. Since both $\mathbb{U}_q(\widetilde{\mathfrak{g}})$ and $\mathcal{H}_q(\mathfrak{S}_n)$ act semisimply on $\mathcal{HT}_{0,q}^{\otimes n}$, the first statement in 1) follows from the bimodule decomposition given in 2). If $N - l \geq n$, then every partition λ of n satisfies the condition $\ell(\lambda) \leq N - l$. It follows that

$$\dim \mathrm{End}_{\mathbb{U}_q}\left(\mathcal{HT}_{0,q}^{\otimes n}\right) = \sum_{\lambda \vdash n} (\dim S(\lambda))^2 = \dim \mathcal{H}_q(\mathfrak{S}_n).$$

Since φ_0 is surjective, this implies that φ_0 must be isomorphism. This proves the second statement in 1).

Let

$$\widehat{y}_{(N-l+1)} := \sum_{w \in \mathfrak{S}_{N-l+1}} (-q)^{-\ell(w)} \widehat{T}_w.$$

We first show that $\widehat{y}_{(N-l+1)} \in \mathrm{Ker}\, \varphi_0$. It suffices to show that for any $\lambda \vdash n$ satisfying $\ell(\lambda) \leq N - l$, $(\widetilde{L}(\lambda) \otimes S(\lambda))\widehat{y}_{(N-l+1)} = 0$.

Since $(\widetilde{L}(\lambda) \otimes S(\lambda)) = \mathbb{U}_q(\widetilde{\mathfrak{g}})z_{0,\lambda}\mathcal{H}_q(\mathfrak{S}_n)$, it is enough to prove that for any $\lambda \vdash n$ satisfying $\ell(\lambda) \leq N - l$ and any $h \in \mathcal{H}_q(\mathfrak{S}_n)$, $z_{0,\lambda}h\widehat{y}_{(N-l+1)} = 0$. Let λ be a partition of n satisfying $\ell(\lambda) \leq N - l$. We write $\mu = (\mu_1, \cdots, \mu_c) = \lambda'$, where $c = \lambda_1$. By the proof of Lemma 3.6 and the fact that $N - l \geq \mu_1 \geq \cdots \geq \mu_c$, we know that

$$z_{0,\lambda} = \Big(v_1 \otimes v_2 \otimes \cdots \otimes v_{\mu_1} \otimes v_1 \otimes v_2 \otimes \cdots \otimes v_{\mu_2} \otimes \cdots \otimes$$
$$v_1 \otimes v_2 \otimes \cdots \otimes v_{\mu_c}\Big)\widehat{y}_{(N-l+1)}.$$

Note that all the vectors v_j in the above simple n-tensor belongs to the subspace of V generated by $\{v_1, v_2, \cdots, v_{N-l}\}$. By a well known result in type A (cf. [14]), we see that $z_{0,\lambda}h\widehat{y}_{(N-l+1)} = 0$, as required.

Therefore, the ideal I generated by $\widehat{y}_{(N-l+1)}$ is contained in $\operatorname{Ker} \varphi_0$. By the work of [23], we know that

$$I = K\text{-Span}\left\{ m_{\mathfrak{s},\mathfrak{t}}^\lambda \,\Big|\, \lambda \vdash n, \lambda_1 > N - l, \mathfrak{s}, \mathfrak{t} \in \operatorname{Std}(\lambda) \right\},$$

where each $m_{\mathfrak{s},\mathfrak{t}}^\lambda$ is the Murphy basis element in [23] with $\widehat{x}_\lambda$ replaced by $\widehat{y}_\lambda$, and $\operatorname{Std}(\lambda)$ denotes the set of standard λ-tableaux with entries in $\{1, 2, \cdots, n\}$. In particular,

$$\dim I = \sum_{\lambda \vdash n, \lambda_1 > N - l} \left(\dim S_\lambda \right)^2 = \sum_{\lambda \vdash n, \ell(\lambda) > N - l} \left(\dim S(\lambda) \right)^2$$

$$= \dim \mathcal{H}_q(\mathfrak{S}_n) - \operatorname{End}_{\mathbb{U}_q(\widehat{\mathfrak{g}})}\left(\mathcal{H}\mathcal{T}_{0,q}^{\otimes n} \right),$$

where S_λ denotes the dual (twist) Specht module of $\mathcal{H}_q(\mathfrak{S}_n)$ associated to λ. Since φ_0 is surjective, this implies that the inclusion $I \subseteq \operatorname{Ker} \varphi_0$ must be an equality. This proves 3).

Corollary 4.2. *With the notations as above, we have that* $V^{\otimes n} y_{(N-l+1)} \subseteq V^{\otimes n} \mathfrak{B}_n^{(1)}$.

Proof. This follows from the isomorphism $V^{\otimes n}/V^{\otimes n}\mathfrak{B}_n^{(1)} \cong \mathcal{H}\mathcal{T}_{0,q}^{\otimes n}$ and Theorem 1.3. $\square$

Corollary 4.3. *Assume that* $m \geq n - 1$. *Then there is a left* $\mathbb{U}_q(\widehat{\mathfrak{g}})$-*module decomposition*

$$V^{\otimes n} = \bigoplus_{f=0}^{[n/2]} m(f) \mathcal{H}\mathcal{T}_{0,q}^{\otimes n - 2f},$$

where the multiplicities are $m(0) = 1$ *and* $m(f) = \binom{n}{2f}(2f - 1)!!$.

Proof. By Theorem 1.2 and 1.3, we have an $\mathbb{U}_q(\widehat{\mathfrak{g}})$-module isomorphism

$$\mathcal{H}\mathcal{T}_{0,q}^{\otimes n - 2f} \cong \bigoplus_{\substack{\lambda \vdash n - 2f \\ \ell(\lambda) \leq N - l}} \dim S(\lambda) \widetilde{L}(\lambda),$$

where $S(\lambda)$ denotes the irreducible $\mathcal{H}_q(\mathfrak{S}_n)$-module corresponding to λ. Now comparing the character formulas of both sides and using the corresponding result [13, Theorem 10.3.3] in the classical case, we prove the corollary. $\square$

Let f be an integer with $0 \le f \le [n/2]$. Let λ be a partition of $n - 2f$. The Young diagram of λ is the set $[\lambda] = \{(i,j) \mid 1 \le j \le \lambda_i\}$. A λ-tableau with entries in $\{2f+1, \cdots, n\}$ is a bijection $\mathfrak{t} \colon [\lambda] \to \{2f+1, 2f+2, \ldots, n\}$. The λ-tableau $\mathfrak{t}$ is standard if $\mathfrak{t}(i,j) \le \mathfrak{t}(i',j')$ whenever $i \le i'$, $j \le j'$. Let $\mathrm{Std}_f(\lambda)$ be the set of standard λ-tableaux with entries in $\{2f+1, \cdots, n\}$. For any $\mathfrak{s} \in \mathrm{Std}_f(\lambda)$, let $d(\mathfrak{s}) \in \mathfrak{S}_{\{2f+1, \cdots, n\}}$ be such that $\mathfrak{t}^\lambda d(\mathfrak{s}) = \mathfrak{s}$.

We set $\nu_f := ((2^f), (n-2f))$, where $(2^f) := \underbrace{(2, 2, \cdots, 2)}_{f \text{ copies}}$ and $(n-2f)$ are considered as partitions of $2f$ and $n-2f$ respectively. So ν_f is a bipartition of n. In a similar way as before, we have the notion of Young diagram for ν_f and (standard) ν_f-tableaux with entries in $\{1, 2, \cdots, n\}$. Let $\mathfrak{t}^{\nu_f}$ be the standard ν_f-bitableau in which the numbers $1, 2, \cdots, n$ appear in order along successive rows of the first component tableau, and then in order along successive rows of the second component tableau. We define

$$\mathcal{D}_{\nu_f} := \left\{ d \in \mathfrak{S}_n \;\middle|\; \begin{array}{l} \mathfrak{t}^{\nu_f} d \text{ is row standard and the first column of } \mathfrak{t}^{(1)} \text{ is an} \\ \text{increasing sequence when read from top to bottom} \end{array} \right\}.$$

For any $\mathfrak{s}, \mathfrak{t} \in \mathrm{Std}_f(\lambda)$, we define

$$m_{\mathfrak{s},\mathfrak{t}} := T_{d(\mathfrak{s})^{-1}} \left(\sum_{w \in \mathfrak{S}_{\{2f+1, \cdots, n\}}} (-q)^{-\ell(w)} T_w \right) T_{d(\mathfrak{t})}.$$

By the main result in [11], we know that the elements in the following set

$$\left\{ T_{d_1^{-1}} E_1 E_3 \cdots E_{2f-1} m_{\mathfrak{s},\mathfrak{t}} T_{d_2} \;\middle|\; \begin{array}{l} 0 \le f \le [n/2], \; \mathfrak{s}, \mathfrak{t} \in \mathrm{Std}_f(\lambda), \; d_1, d_2 \in \mathcal{D}_{\nu_f}, \\ \text{where } \lambda \vdash n - 2f, \; \nu_f := ((2^f), (n-2f)) \end{array} \right\}$$

is a cellular basis of the specialized BMW algebra $\mathfrak{B}_{n,q}$.

For any two partitions λ, μ of $n - 2f$, we write $\lambda \trianglerighteq \mu$ if $\sum_{j=1}^{i} \lambda_j \ge \sum_{j=1}^{i} \mu_j$, for all $i \ge 1$. If f, g are two integers with $0 \le f, g \le [n/2]$ and $\lambda \vdash n - 2f, \mu \vdash n - 2g$, then we write $(f, \lambda) \trianglerighteq (g, \mu)$ if either $f > g$ or $f = g$ and $\lambda \trianglerighteq \mu$. We write $(f, \lambda) \triangleright (g, \mu)$ if $(f, \lambda) \trianglerighteq (g, \mu)$ and $(f, \lambda) \ne (g, \mu)$.

For each integer f with $0 \le f \le [n/2]$ and each $\lambda \vdash n - 2f$, we use $\mathfrak{B}_{n,q}^{\triangleright(f,\lambda)}$ the K-subspace of $\mathfrak{B}_{n,q}$ spanned by all the basis elements in the following set:

$$\left\{ T_{d_1^{-1}} E_1 E_3 \cdots E_{2g-1} m_{\mathfrak{s},\mathfrak{t}} T_{d_2} \;\middle|\; \begin{array}{l} f \le g \le [n/2], \; \mathfrak{s}, \mathfrak{t} \in \mathrm{Std}_g(\mu), \; d_1, d_2 \in \mathcal{D}_{\nu_g}, \\ \text{where } \mu \vdash n - 2g, \; \nu_g := ((2^g), (n-2g)) \\ (g, \mu) \triangleright (f, \lambda) \end{array} \right\}.$$

By the cellular structure of $\mathfrak{B}_{n,q}$, we know that $\mathfrak{B}_{n,q}^{\triangleright(f,\lambda)}$ is a two–sided ideal of $\mathfrak{B}_{n,q}$.

We use $S(f, \lambda)$ to denote the cell module associated to (f, λ). Since q is generic, $\mathfrak{B}_{n,q}$ is also quasi-hereditary. By general theory of quasi-hereditary cellular algebra (cf. [12]), we know that each $S(f, \lambda)$ has a unique simple head $D(f, \lambda)$. We are now in a position to give a proof of Theorem 1.4.

Proof of Theorem 1.4: We only give the proof in the case when $X \in \{B, C\}$. The proof for the case $X = D$ is similar. By definition, $S(f, \lambda')$ is the right $\mathfrak{B}_{n,q}$-submodule of $\mathfrak{B}_{n,q}/\mathfrak{B}_{n,q}^{\triangleright(f,\lambda')}$ generated by

$$m_{f,\lambda'} := E_1 E_3 \cdots E_{2f-1} y_{\lambda'} + \mathfrak{B}_{n,q}^{\triangleright(f,\lambda')}.$$

We claim that the map τ which sends $m_{f,\lambda'}b + \mathfrak{B}_{n,q}^{\triangleright(f,\lambda')}$ (for each $b \in \mathfrak{B}_{n,q}$) to $z_{f,\lambda}b$ gives a surjective right $\mathfrak{B}_{n,q}$-module homomorphism from $S(f, \lambda')$ onto $z_{f,\lambda}\mathfrak{B}_{n,q}$.

Indeed, it suffices to show that τ is well-defined. To this end, we assume that $m_{f,\lambda'}b + \mathfrak{B}_{n,q}^{\triangleright(f,\lambda')} = \mathfrak{B}_{n,q}^{\triangleright(f,\lambda')}$, i.e., $E_1 E_3 \cdots E_{2f-1} y_{\lambda'} b \in \mathfrak{B}_{n,q}^{\triangleright(f,\lambda')}$. By the property (C3) (cf. [12, Definition (1.1)]) of cellular basis, we deduce that

$$E_1 E_3 \cdots E_{2f-1} y_{\lambda'} b \equiv \sum_{\substack{\mathbf{s},\mathbf{t}\in \mathrm{Std}_f(\mu) \\ \mu \triangleright \lambda', d \in \mathcal{D}_{\nu_f}}} r_{\mathbf{s},\mathbf{t}} E_1 E_3 \cdots E_{2f-1} m_{\mathbf{s},\mathbf{t}} T_d \pmod{\mathfrak{B}_n^{(f+1)}},$$

where $r_{\mathbf{s},\mathbf{t}} \in K$ for each $\mathbf{s}, \mathbf{t}$. By our assumption on $\ell(\lambda)$, we see that $z_{f,\lambda}\mathfrak{B}_n^{(f+1)} = 0$. Therefore,

$$
\begin{aligned}
z_{f,\lambda}b &= \left(\alpha^{\otimes f} \otimes v_\lambda T_{w_\lambda} y_{\lambda'}\right)b \\
&= x^{-f}\left(\alpha^{\otimes f} \otimes v_\lambda T_{w_\lambda}\right)E_1 E_3 \cdots E_{2f-1} y_{\lambda'} b \\
&= x^{-f}\left(\alpha^{\otimes f} \otimes v_\lambda T_{w_\lambda}\right) \sum_{\substack{\mathbf{s},\mathbf{t}\in \mathrm{Std}_f(\mu) \\ \mu \triangleright \lambda', d \in \mathcal{D}_{\nu_f}}} r_{\mathbf{s},\mathbf{t}} E_1 E_3 \cdots E_{2f-1} m_{\mathbf{s},\mathbf{t}} T_d \\
&= \left(\alpha^{\otimes f} \otimes v_\lambda T_{w_\lambda}\right) \sum_{\substack{\mathbf{s},\mathbf{t}\in \mathrm{Std}_f(\mu) \\ \mu \triangleright \lambda', d \in \mathcal{D}_{\nu_f}}} r_{\mathbf{s},\mathbf{t}} m_{\mathbf{s},\mathbf{t}} T_d \\
&= \alpha^{\otimes f} \otimes \sum_{\substack{\mathbf{s},\mathbf{t}\in \mathrm{Std}_f(\mu) \\ \mu \triangleright \lambda', d \in \mathcal{D}_{\nu_f}}} r_{\mathbf{s},\mathbf{t}} v_\lambda T_{w_\lambda} m_{\mathbf{s},\mathbf{t}} T_d \\
&= \alpha^{\otimes f} \otimes \sum_{\substack{\mathbf{s},\mathbf{t}\in \mathrm{Std}_f(\mu) \\ \mu \triangleright \lambda', d \in \mathcal{D}_{\nu_f}}} r_{\mathbf{s},\mathbf{t}} v_\lambda \widehat{T}_{w_\lambda} \widehat{m}_{\mathbf{s},\mathbf{t}} \widehat{T}_d,
\end{aligned}
$$

where $\widehat{m}_{s,t} := \widehat{T}_{d(s)^{-1}}\widehat{y}_\mu \widehat{T}_{d(t)}$ is the Murphy basis element in the Hecke algebra $\mathcal{H}_q(\mathfrak{S}_{\{2f+1,\cdots,n\}})$, and we have used our assumption on $\ell(\lambda)$ in the last step.

On the other hand, by [9, (4.1)], we know that $\widehat{x}_\lambda \widehat{T}_{w_\lambda} \widehat{m}_{s,t} = 0$ whenever $\mu \rhd \lambda'$. It follows that

$$v_\lambda \widehat{T}_{w_\lambda} \widehat{m}_{s,t} = \Big(\sum_{w \in \mathfrak{S}_\lambda} q^{\ell(w)} \Big)^{-1} v_\lambda \widehat{x}_\lambda \widehat{T}_{w_\lambda} \widehat{m}_{s,t} = 0.$$

As a result, we have that $z_{f,\lambda} b = 0$, as required. This proves our claim.

Now by our assumption on X and $\ell(\lambda)$, we know that $z_{f,\lambda} \mathfrak{B}_{n,q} \cong \widetilde{D}_{f,\lambda}$ is a simple $\mathfrak{B}_{n,q}$-module. Since $S(f,\lambda')$ has a unique simple head $D(f,\lambda')$, the surjectivity of τ immediately implies that $D(f,\lambda') \cong z_{f,\lambda} \mathfrak{B}_{n,q}$ as required. This completes the proof of Theorem 1.4.

Acknowledgments

The research was supported by an Australian Research Council discovery grant, the National Natural Science Foundation of China (Project 10771014), the Basic Research Foundation of BIT and the Scientific Research Foundation for the Returned Overseas Chinese Scholars, State Education Ministry.

References

1. J. Birman and H. Wenzl, Braids, link polynomials and a new algebra, *Trans. Amer. Math. Soc.* **313**(1) (1989) 249–273.
2. R. Brauer, On algebras which are connected with semisimple continuous groups, *Ann. Math.* **38** (1937) 857–872.
3. W. P. Brown, An algebra related to the orthogonal group, *Michigan Math. J.* **3** (1955–1956) 1–22.
4. W. P. Brown, The semisimplicity of ω_f^n, *Ann. Math.* **63** (1956) 324–335.
5. V. Chari and A. Pressley, *A guide to quantum groups*, Cambridge University Press, Cambridge, 1994.
6. C. De Concini and C. Procesi, A characteristic free approach to invariant theory, *Adv. Math.* **21** (1976) 330–354.
7. R. Dipper, S. Doty and J. Hu, Brauer algebras, symplectic Schur algebras and Schur-Weyl duality, *Trans. Amer. Math. Soc.* **360**(1) (2008) 189–213.
8. S. Doty and J. Hu, Schur–Weyl duality for orthogonal groups, *Proceedings of the London Mathematical Society* **98** (2009) 679–713.
9. R. Dipper and G. D. James, Representations of Hecke algebras of general linear groups, *Proc. London. Math. Soc.* **52**(3) (1986) 20–52.

10. R. Dipper and G. D. James, q-tensor spaces and q-Weyl modules, *Trans. Amer. Math. Soc.* **327**(1) (1991) 251–282.

11. J. Enyang, Cellular bases for the Brauer and Birman–Murakami–Wenzl algebras, *J. Algebra* **281** (2004) 413–449.

12. J. J. Graham and G. I. Lehrer, Cellular algebras, *Invent. Math.* **123** (1996) 1–34.

13. R. Goodman and N. R. Wallach, *Representations and invariants of classical groups*, Cambridge University Press, 1998.

14. M. Härterich, Murphy bases of generalized Temperley-Lieb algebras, *Arch. Math.* **72**(5) (1999) 337–345.

15. T. Hayashi, Quantum deformation of classical groups, *Publ. RIMS. Kyoto Univ.* **28** (1992) 57–81.

16. J. Hu and Z. Li, On tensor spaces over Hecke algebras of type B_n, *J. Algebra* **304** (2006) 602–611.

17. J. Hu, BMW algebra, Quantized coordinate algebra and type C Schur–Weyl duality, Representation Theory, an electronic journal of the AMS, article in press, arXiv:0708.3009.

18. J. Hu, Dual partially harmonic tensors and Brauer–Schur–Weyl duality, *Transformation Groups* **15** (2010) 333–370.

19. J. Hu and Z. Xiao, On tensor spaces for Birman–Murakami–Wenzl algebras, *J. Algebra* **324** (2010) 2893–2922.

20. J. Hu and Y. Yang, Some irreducible representations of Brauer's centralizer algebras, *Glasgow Mathematical Journal* **46** (2004) 499–513.

21. Q. Liu, Schur algebras of classical groups, *J. Algebra* **301** (2006) 867–887.

22. J. Murakami, The Kauffman polynomial of links and representation theory, *Osaka J. Math.* **26**(4) (1987) 745–758.

23. E. Murphy, The representations of Hecke algebras of type A_n, *J. Algebra* **173** (1995) 97–121.

24. H. Wenzl, On the structure of Brauer's centralizer algebras, *Ann. Math.* **128** (1988) 173–193.

QUANTUM ENTANGLEMENT AND APPROXIMATION BY POSITIVE MATRICES

XIAOFEN HUANG

School of Mathematics and Statistics, Hainan Normal University,
Haikou, Hainan 571158, China
E-mail: huangxf1206@googlemail.com

NAIHUAN JING

Department of Mathematics, North Carolina State University,
Raleigh, NC 27695-8205, USA
and
School of Sciences, South China University of Technology
Guangzhou, Guangdong 510640, China
E-mail: jing@math.ncsu.edu

We give an exact solution to the nonlinear optimization problem of approximating a Hermitian matrix by positive semi-definite matrices. Our algorithm was then used to judge whether a quantum state is entangled or not. We show that the exact approximation of a density matrices by tensor product of positive semi-definite operators is determined by the additivity property of the density matrix.

Keywords: Quantum computation; density matrices; nonlinear optimization.

1. Introduction

Quantum entanglement is one of the most interesting features in quantum computation[20,23] as it is vital for quantum dense coding, quantum error-correcting codes, teleportation and also responsible for fast quantum algorithms. Much efforts have been made to find criteria for quantum separability: the Bell inequalities,[3] PPT (positive partial transposition),[23] reduction criterion,[5,12] majorization criterion,[22] entanglement witnesses,[16,19,26] realignment[6,24] and generalized realignment methods.[1] Nevertheless the problem is still not fully solved except for lower rank cases.[2,7,16]

In[8] we proposed a new method to judge if a density matrix is separable. We first decompose the density matrix as a tensor product of hermitian matrices, and then we reduce the separability problem to that of finding when the hermitian matrix becomes positive semi-definite. The strategy was to solve the separability by two steps: first one finds a tensor decomposition of the density matrix by hermitian matrices, and then one approximates the hermitian matrices by positive semi-definite matrices if possible. Although it was proved that a density matrix is separable if and only if the separability indicator is non-negative, it is highly nontrivial to actually compute this indicator. In this sense the method is also similar to many of its predecessors.

In the current paper we approach the question from a new angle by giving an algorithm to compute the closest positive semi-definite matrix to any given hermitian matrix. It is noted that this optimization is not a linear problem so the usual QR decomposition in[15] does not work. By general theory of convex sets the existence of the minimum is guaranteed but it is nontrivial to find the exact solution due to the nonlinearity. In this paper we first solve this optimization problem exactly using matrix theory. This paves the way for us to attack the main problem of separability by directly looking for an optimal approximation to a sum of matrices by positive semi-definite matrices.

It turns out that the exact solution in the most important case can be solved by Lie theoretic techniques. First we show that a Hermitian matrix with two commuting summands can be approximated term by term. Next we prove that if the summands of a Hermitian matrix can be simultaneously made to upper triangular matrices, then the exact approximation by positive semi-definite matrix can be done term by term. Our method provides a direct way to approximate the hermitian matrix by positive semi-definite matrices and thus the separability problem can be solved theoretically in this sense.

It is interesting to note that similar (but stronger constrained) matrix approximation also appears in finance, image processing, date mining, and other areas such as resource allocation and industrial process monitoring.[4,25,28] Most methods used in these problems are numerical algorithms that only give an approximation to the solution. In some sense our results also provide the first exact and analytical method in this direction, and we also hope that conversely some of the numerical algorithms may be useful in quantum entanglements.

2. Approximation by Positive Definite Matrices

Let A be an $n \times n$ Hermitian matrix, and let Q be a unitary matrix such that $A = QDQ^\dagger$, where $D = diag(\alpha_1, \cdots, \alpha_n)$ and $\dagger$ is conjugation and transposition. The signature (p, q) of A (cf.[18]) is defined by $p+q = rank(A)$. We can permute the columns of the matrix Q so that the eigenvalues of A are arranged in the following order:

$$\alpha_1 \geq \cdots \geq \alpha_p > 0 > \alpha_{p+1} \geq \cdots \geq \alpha_{p+q}, \alpha_{p+q+1} = \cdots = \alpha_n = 0.$$

We further define

$$A = A_+ - A_-, \tag{1}$$

where $A_\pm = QD_\pm Q^\dagger$, and

$$D_+ = diag(\alpha_1, \cdots, \alpha_p, 0, \cdots, 0)$$
$$D_- = diag(0, \cdots, 0, -\alpha_{p+1}, \cdots, -\alpha_{p+q}, 0, \cdots, 0)$$

formed by positive (negative) eigenvalues respectively. We remark that our definition of the positive and negative semi-definite parts of A is independent from our choice of Q. In general, if A is diagonalized by a unitary matrix Q as follows:

$$A = QDQ^\dagger = Qdiag(\alpha_1, \cdots, \alpha_n)Q^\dagger,$$

then we have

$$A_\pm = Qdiag(\alpha_1^\pm, \cdots, \alpha_n^\pm)Q^\dagger, \qquad \alpha_i^\pm = \frac{|\alpha_i| \pm \alpha_i}{2}.$$

It is clear that A is positive semi-definite if and only all eigenvalues of A are non-negative. Therefore both $A_\pm$ are positive semi-definite matrices. It is easy to see that the decomposition (1) of A into a difference of positive semi-definite matrices is unique up to positive definite matrices, i.e., if $A = A'_+ - A'_-$, where $A'_\pm$ are positive semi-definite, then $A_\pm = A'_\pm + P$ with a positive semi-definite matrix P.

If A and B are commuting positive semi-definite matrices, so is their product. If they are not commutative, then AB is in general not a positive semi-definite matrix, as AB may not even be hermitian.

Lemma 2.1. *Let A and B be positive semi-definite Hermitian matrices, then the eigenvalues of AB are all non-negative.*

Proof. If A is invertible, and let λ be an eigenvalue of AB, then for some vector $x \neq 0$,

$$ABx = \lambda x.$$

Hence $Bx = \lambda A^{-1}x$. Note that both A^{-1} and B are positive, so

$$x^\dagger Bx = \lambda x^\dagger A^{-1}x > 0,$$

Subsequently $\lambda > 0$. In general, if A is singular, then for any $\epsilon > 0$ the matrix $A + \epsilon I$ is positive definite. Therefore any eigenvalue $\lambda = \lambda(\epsilon)$ of $(A + \epsilon I)B$ is positive. Letting $\epsilon \to 0$, we see that $lim_{\epsilon \to 0}\lambda \geq 0$, i.e. any eigenvalue of AB is non-negative. □

For any matrix A, the Frobenius norm is defined to be $\|A\|_F = (tr(AA^\dagger))^{1/2}$, which is also equal to the sum of squares of singular values of A.

Theorem 2.1. *Let A be an $n \times n$ hermitian matrix, then for any positive semi-definite matrix B we have*

$$\|A - B\|_F \geq \|A_-\|_F \tag{2}$$

with equality when $B = A_+$. i.e. the closest positive semi-definite matrix to A is given by A_+.

Proof. Omitting the subscript in the Frobenius norm, we have for any positive semi-definite matrix B

$$\begin{aligned}
\|A - B\|^2 &= tr(A - B)^2 = tr(A^2) - 2tr(AB) + tr(B^2) \\
&= tr(A^2) + 2tr(A_-B) + tr(B^2) - 2tr(A_+B) \\
&= tr(A^2) + 2tr(A_-B) + tr(B - A_+)^2 - tr(A_+^2)
\end{aligned}$$

by using $tr(AB) = tr(BA)$ and completing square. Since B and $A_\pm$ are positive semi-definite, then $tr(A_-B) \geq 0$ for any B by Lemma 2.1. Therefore, it follows that for any positive semi-definite matrix B

$$\begin{aligned}
\|A - B\|^2 &\geq tr(A^2) + tr(B - A_+)^2 - tr(A_+^2) \\
&= \|A\|^2 - \|A_+\|^2 + \|B - A_+\|^2 \\
&\geq \|A\|^2 - \|A_+\|^2
\end{aligned}$$

where the equality is obtained when $B = A_+$. □

We remark that similar (Toeplitz and/or correlation matrix) approximation with stronger constraints has been studied in finance and image processing.[25,28] Our result is more general and stronger in the sense that we do not require that the matrix to be either Toeplitz or correlation matrix (real positive definite with unit diagonal). Furthermore our result is analytical and exact, as no numerical approximation is needed for the solution.

Corollary 2.1. *Let A and B be any two Hermitian matrices, then for any positive semi-definite matrix C of the same size as $A \otimes B$, we have*

$$\|A \otimes B - C\|_F \geq \|(A \otimes B)_-\|_F = \|A_+ \otimes B_- + A_- \otimes B_+\|_F.$$

The equality holds when $C = (A \otimes B)_+ = A_+ \otimes B_+ + A_- \otimes B_-$.

Proof. It is enough to show that $(A \otimes B)_\pm = \sum_\epsilon A_\epsilon \otimes B_{\pm\epsilon}$, where $\epsilon = +, -$. Suppose A and B are diagonalized by Q_1 and Q_2 respectively:

$$A = Q_1 D_1 Q_1^\dagger = Q_1 diag(\alpha_1, \cdots, \alpha_m) Q_1^\dagger,$$
$$B = Q_2 D_2 Q_2^\dagger = Q_2 diag(\beta_1, \cdots, \beta_n) Q_2^\dagger,$$

then we have

$$A \otimes B = (Q_1 \otimes Q_2)(D_1 \otimes D_2)(Q_1 \otimes Q_2)^\dagger$$

As the eigenvalues σ of $A \otimes B$ are $\lambda_i \beta_j$, thus

$$\sigma_{ij}^+ = \alpha_i^+ \beta_j^+ + \alpha_i^- \beta_j^-,$$
$$\sigma_{ij}^- = \alpha_i^+ \beta_j^- + \alpha_i^- \beta_j^+,$$
$$\sigma_{ij}^0 = \alpha_i^0 \beta_j + \alpha_i \beta_j^0.$$

Since the zero eigenvalues do not contribute to the decomposition, we have that $(A \otimes B)_+ = A_+ \otimes B_+ + A_- \otimes B_-$, and $(A \otimes B)_- = A_+ \otimes B_- + A_- \otimes B$ □

3. Sums of Matrices and Estimates

It appears that the decomposition of the Hermitian matrix A into a sum of two Hermitian matrices B, C has a close relationship to our problem. This problem has a much longer history in mathematics.

Horn[11] defines the following concept for the Hermitian matrices. Let A and B be two $n \times n$ matrices with eigenvalues α_i and β_i. For any subset I we denote $|I| = \sum_{i \in I} i$. The set T_r^n of triples (I, J, K) of subsets of $\{1, \ldots, n\}$ of the same cardinality r is defined by first setting

$$U_r^n = \{(I, J, K) | |I| + |J| = |K| + r(r+1)/2\},$$

then define $T_1^n = U_1^n$ and in general

$$T_r^n = \{(I, J, K) \in U_r^n | \text{ for all } p < r \text{ and all } (F, G, H) \in U_p^n$$
$$\sum_{f \in F} i_f + \sum_{g \in G} j_g \leq \sum_{h \in H} k_h + p(p+1)/2\},$$

The following characterization of eigenvalues α, β, γ of $C = A + B$ is proved in.[9]

Theorem 3.1. *(Horn's conjecture)* *A triple* (α, β, γ) *occurs as eigenvalues of* A, B, C *such that* $C = A + B$ *if and only if* $|\alpha| + |\beta| = |\gamma|$ *and inequalities*

$$\sum_{k \in K} \gamma_k \leq \sum_{i \in I} \alpha_i + \sum_{j \in J} \beta_j$$

hold for all triple (I, J, K) *in* T_r^n *for all* $r < n$.

The special case of $r = 1$ is Weyl's inequality:

$$\gamma_{i+j-1} \leq \alpha_i + \beta_j$$

Since $|\gamma| = |\alpha| + |\beta|$, we also have

$$\sum_{k \in K^c} \gamma_k \geq \sum_{i \in I^c} \alpha_i + \sum_{j \in J^c} \beta_j$$

A practical bound is the following:

$$\operatorname*{Max}_{i+j=n+k} \alpha_i + \beta_j \leq \gamma_k \leq \operatorname*{Min}_{i+j=k+1} \alpha_i + \beta_j$$

Apply these results to our situation, we then get the following:

Theorem 3.2. *Let* A *be a density matrix over the Hilbert space* $\mathcal{H} = \mathcal{H}_1 \otimes \mathcal{H}_2$. *Suppose* $A = \sum_i B_i \otimes C_i$, *then we have*

$$\|A - A_+\|_F \leq \sum_i \|B_i - (B_i)_+\| \cdot \|C_i - (C_i)_+\|$$

4. Lie Algebras and Approximation of Summations

As the Hermitian decomposition of the density matrix A is a summation, one hopes to check its separability by demonstrating that each summand can be expressed by tensor products of positive semi-definite matrices.

The main problem is to estimate the error given by term by term approximation. Suppose that A and B are two Hermitian matrices we would like to estimate the norm $\|A + B - A_+ - B_+\|$.

Lemma 4.1. *Let* A *and* B *are two commuting Hermitian matrices, then*

$$(A + B)_+ \leq A_+ + B_+$$
$$(A + B)_- \leq A_- + B_-$$

Proof. Since A and B are commuting, they can be diagonalized simultaneously by a unitary matrix Q. In other words we have

$$A = QD_AQ^\dagger = A_+ - A_- = Q(D_A)_+Q^\dagger - Q(D_A)_-Q^\dagger$$
$$B = QD_BQ^\dagger = B_+ - B_- = Q(D_B)_+Q^\dagger - Q(D_B)_-Q^\dagger$$

thus,

$$A + B = Q(D_A + Q_B)Q^\dagger$$
$$= Q((D_A)_+ + (D_B)_+)Q^\dagger - Q((D_A)_- + (D_B)_-)Q^\dagger,$$

which implies that $(A + B)_+ = A_+ + B_+$ and $(A + B)_- = A_- + B_-$. $\square$

Theorem 4.1. *Let A_i (resp. B_i) be set of commuting positive semi-definite Hermitian matrices of the same size, then for any positive definite matrix C of the same size as $A_i \otimes B_i$ we have*

$$\left\| \sum_i A_i \otimes B_i - C \right\|_F \geq \left\| \sum_i \{(A_i)_+ \otimes (B_i)_- + (A_i)_- \otimes (B_i)_+\} \right\|_F,$$

where the equality holds when

$$C = \sum_i (A_i \otimes B_i)_+ = \sum_i \{(A_i)_+ \otimes (B_i)_+ + (A_i)_- \otimes (B_i)_-\}.$$

We remark that the condition that A and B are commuting with each other is also necessary for the equality in the theorem to hold. The following result is quoted from standard books on Lie algebras.[10]

Proposition 4.1. *(Lie's theorem) Let L be any solvable subalgebra of the general linear Lie algebra, then the matrices of L relative to a suitable basis of V are upper triangular. Furthermore, one can adjust the basis to be orthonormal.*

We remark that the last statement is due to the fact that the transition matrix in Gram-Schmidt process is upper triangular. Now we suppose that there exists a unitary matrix Q such that both A and B are upper-triangularized as follows:

$$A = Q\Lambda_1Q^\dagger, B = Q\Lambda_2Q^\dagger,$$

where Λ_i are upper-triangular. If A and B are hermitian, then $\Lambda_i^\dagger = \Lambda_i$, which forces Λ_i to be diagonal. Then A and B are actually commuting with each other, subsequently $(A + B)_\pm = A_\pm + B_\pm$. So in this context the additivity seems to be not too far away from commutativity.

Theorem 4.1 gives the closest approximation to a two-partite density operator by the tensor operator of non-negative operators, however one has to fit the approximation under the constraint of unit trace. We hope that further studies can answer this question.

5. Conclusion

Matrix approximation is an old problem in mathematics with applications in physics, finance, and computer sciences. In this paper we have completely solved the optimization problem to approximate any Hermitian matrix by positive semi-definite matrices. The solution is shown to be given by the spectral decomposition of the concerned matrix. We apply this result to density matrices and obtain useful approximation by tensor product of density matrices using Lie theoretic techniques. Our results also open possible deep connection among quantum entanglement, data mining and signal procession.

Acknowledgments

Jing gratefully acknowledges the support from NSA grant MDA904-97-1-0062 and NSFC's Overseas Distinguished Youth Grant (10801094).

References

1. S. Albeverio, K. Chen and S. M. Fei, Phys. Rev. A 68, 062313 (2003).
2. S. Albeverio, S. M. Fei and D. Goswami, Phys. Lett. A 286, 91 (2001).
3. J. S. Bell, J. Physics (N.Y.) 1, 195 (1964).
4. M.-A. Belabbas, P. J. Wolfe, *On the approximation of matrix products and positive definite matrices*, arxiv:0707.4448 (2007).
5. N. J. Cerf, C. Adami and R. M. Gingrich, Phys. Rev. A 60, 898 (1999).
6. K. Chen and L. A. Wu, Quant. Inf. Comput. 3, 193 (2003); K. Chen and L. A. Wu, Phys. Lett. A 306, 14 (2002).
7. S. M. Fei, X. H. Gao, X. H. Wang, Z. X. Wang and K. Wu, Phys. Lett. A 300, 555 (2002).
8. S. Fei, N. Jing, B. Sun, Rep. Math. Phys. **57** (2006), no. 2, 271–288.
9. W. Fulton, Bull. Amer. Math. Soc. **37**, 209 (2000).
10. J. E. Humphreys, *Introduction to Lie algebras and representation theory*, Springer-Verlag, New York, 1970.
11. A. Horn, Pac. J. Math. 12 (1962) 225–241.
12. M. Horodecki, P. Horodecki, Phys. Rev. A 59, 4206 (1999).
13. M. Horodecki, P. Horodecki and R. Horodecki, Phys. Lett. A 223, 1 (1996).
14. X. Huang and N. Jing, *Separability of Multi-partite quantum states*, J. Phys. A, to appear, arXiv:0807.5003 (2008).

15. R. A. Horn and C. R. Johnson, *Matrix Analysis*, Cambridge University Press, Cambridge, U.K., 1985.
16. P. Horodecki, M. Lewenstein, G. Vidal and I. Cirac, Phys. Rev. A 62, 032310 (2000).
17. L. P. Hughston, R. Jozsa, and W. K. Wooters, Phys. Lett. A 183, 14 (1993).
18. S. Lang, *Linear Algebra*, 3rd ed., Springer-Verlag, New York, 1987.
19. M. Lewenstein, B. Kraus, J. I. Cirac and P. Horodecki, Phys. Rev. A 62, 052310 (2000).
20. M. A. Nielsen and I. L. Chuang, *Quantum computation and quantum information*, Cambridge Univ. Press, 2000.
21. M. A. Nielsen, Phys. Rev. A **62**, 052308 (2000)
22. M. A. Nielsen and J. Kempe, Phys. Rev. Lett. 86, 5184 (2001).
23. A. Peres, Phys. Rev. Lett. 77, 1413 (1996).
24. O. Rudolph, Phys. Rev. A 67, 032312 (2003).
25. T. J. Suffridge, T. L. Hayden, SIAM J. Matrix Anal. Appl. 14, 721 (1993).
26. B. Terhal, Phys. Lett. A 271, 319 (2000).
27. H. Weyl, Math. Ann. **71**, 441 (1912).
28. Z. Zhang, L. Wu, Linear Algebra Appl. 364, 161 (2003).

2-PARTITIONS OF ROOT SYSTEMS

BIN LI

School of Mathematics and Statistics, Wuhan University,
Wuhan, 430072, P. R. China
E-mail: li_bin789@yahoo.com.cn

WILLIAM WONG

Department of Pure Mathematics and Mathematical statistics, University of
Cambridge, UK
E-mail: hw340@cam.ac.uk

HECHUN ZHANG

Department of Mathematical Science, Tsinghua University,
Beijing, 100084, P. R. China
E-mail: hzhang@math.tsinghua.edu.cn

The 2-partitions of the set of positive roots and of the root system of finite type or affine type are completely classified.

Keywords: Root system; Weyl group; additionally closed; 2-partition.

1. Introduction

Root systems were introduced by W. Killing and E. Cartan for the study of semisimple Lie algebras and now they are basic in several fields of mathematics.

A subset A of a root system R is called additionally closed if for any $\alpha, \beta \in A$, $\alpha + \beta \in R$ implies $\alpha + \beta \in A$. This kind of additionally closed subsets have been studied in various literature, e.g. Ref. 1–5. In the present article, we are interested in the so-called partitions of root systems of finite type and respectively, of these sets of positive roots, namely, a collection of pairwise disjoint additionally closed subsets $S_1, S_2, \cdots, S_l$ of the root system R (resp. R_+) such that $R = S_1 \cup S_2 \cup \cdots \cup S_l$ (resp. $R_+ = S_1 \cup S_2 \cup \cdots \cup S_l$).

Among all of the partitions, the 2-partitions are of particular interests.

In order to define root vectors of quantum groups using Lusztig's automorphisms, one need to fix a convex order on the set of positive roots R_+ which is equivalent to fix a reduced expression of the longest element in the Weyl group. The 2-partitions of the set of positive roots R_+ are related to the convex orders on R_+ which were classified in Ref. 2 (see also Refs. 1,3,4 for the convex order on the set of positive roots of affine type). A special kind of 2-partitions of affine type root systems are determined in Refs. 6,7 which is used in the construction of some unitary modules and generalized Verma modules over affine Kac-Moody algebras. It seems that the following assumptions about the representations of Kac-Moody algebra and associated quantum groups are natural and fundamental:

(1) The representation is a highest weight module with respect to a properly chosen 2-partitions of the root system.
(2) There is a non-degenerate contravariant bilinear form on the representation.

In the present paper, we only deal with the first assumption on representations. The 2-partitions of the root systems of finite type or affine type are completely classified.

The arrangement of the paper is as follows. In section 2, we recall some basics of root systems which will be used in the sequel and fix some notations. In section 3, we classify the 2-partitions of the set of positive roots as it has been done in Ref. 2. However, it seems that our proof is simpler. In sections 4 and 5, the 2-partitions of the root system of finite type or affine type are completely determined. These are the main results of the paper.

2. Preliminaries

In this section we review root systems and some related notations (see, for example Ref. 8). Fix a standard inner product $(\ ,\ )$ of an Euclidean space E and the orthonormal basis $\{\epsilon_1, \ldots, \epsilon_n\}$ of E. For $\alpha \in E \setminus \{0\}$, the *reflection* s_α *with respect to* α is defined by

$$s_\alpha : E \longrightarrow E$$
$$x \longmapsto x - 2\frac{(\alpha,x)}{(\alpha,\alpha)}\alpha$$

Definition 2.1. A (reduced) **root system of finite type** is a set R of vectors in an Euclidean space E satisfying the following axioms:

(1) R is finite, spans E and does not contain 0.
(2) The only multiples in R of a vector $\alpha \in R$ are $\pm\alpha$.

(3) For all $\alpha \in R$, the reflection s_α with respect to α leaves R invariant.

(4) For all $\alpha, \beta \in R$, $\frac{2(\alpha,\beta)}{(\beta,\beta)} \in \mathbb{Z}$.

For the sake of simplicity, we write $\frac{2(\alpha,\beta)}{(\beta,\beta)}$ as $\langle \alpha, \beta \rangle$, for $\alpha, \beta \in R$.

In the present paper, we are mainly interested in the subsets of a root system which are "closed" under addition.

Definition 2.2. A subset A of a root system R is said to be **additionally closed**, or closed for short if

$$\alpha, \beta \in A, \ \alpha + \beta \in R \Rightarrow \alpha + \beta \in A.$$

The basis of a root system will play a key role in our discussion.

Definition 2.3. A subset Π of the root system R is called a basis for R if the following two conditions are satisfied:

(1) Π is a basis of the underlying vector space E.

(2) Each root $\beta \in R$ can be written as

$$\beta = \sum_{\alpha \in \Pi} m_\alpha \alpha,$$

where the coefficients m_α are integers with the same sign (i.e. all ≥ 0 or ≤ 0).

It is well-known that any root system R has at least one basis and all bases are conjugate under the action of the Weyl group (see Ref. 8). From now on we shall denote by Π a fixed basis of R. Denote $\Pi = \{\alpha_1, \alpha_2, \cdots, \alpha_n\}$. The roots in Π are called simple roots. The dimension $dim E = |\Pi|$ of E will be called the rank of R and denoted by $rank(R)$.

Let $Q = \sum_{\alpha \in \Pi} \mathbb{Z}\alpha$ be the root lattice and let $Q_+ = \sum_{\alpha \in \Pi} \mathbb{Z}_+\alpha$ be the positive root lattice. The subsets $R_+ := Q_+ \cap R$ and $R_- := -Q_+ \cap R$ are called the set of positive roots and the set of negative roots respectively. Clearly, $R = R_+ \cup R_-, R_+ = -R_-$.

Denote by $A = (\langle \alpha_i, \alpha_j \rangle)_{i,j}$ the Cartan matrix of R and by $D(A)$ the Dynkin diagram of A. Let W be the Weyl group of R. Denote by s_i the reflection with respect to the simple root α_i. It is known that the Weyl group W is generated by the reflections s_i for all $1 \leq i \leq n$. For a subset $T \subset \Pi$, denote by $\langle T \rangle$ the subspace of E spanned by T. It is easy to check that $R(T) = \langle T \rangle \cap R$ is a sub root system which is called the root system generated by T.

For any $\beta = \sum_i m_i \alpha_i \in Q$, the number $ht\beta := \sum_i m_i$ is called the height of β. We define the support of β (written as $supp\beta$) to be the sub-diagram of the Dynkin diagram $D(A)$ which consists of those vertices i for which $m_i \neq 0$ and of all the lines joining these vertices. It is known that $supp\beta$ is connected if $\beta \in R$.

The following lemma is well-known.[8]

Lemma 2.1. *For any simple root $\alpha_i \in \Pi$, $s_i(R_+) = (R_+ \setminus \{\alpha\}) \cup \{-\alpha\}$.*

3. 2-partitions of the root system of finite type

For a given $w \in W$, write

$$N(w) = \{\alpha \in R_+ | w(\alpha) \in R_-\}$$

for the set of all positive roots that are mapped to negative roots by w.

Remark 3.1. $N(w)$ is a closed set for all $w \in W$.

The following proposition is well known.

Proposition 3.1. $\mid N(w) \mid = l(w)$ *the length of w in the Weyl group.*

Definition 3.1. A 2-partition of R_+ is a pair of additionally closed subsets S_1, S_2 satisfying

$$R_+ = S_1 \cup S_2, S_1 \cap S_2 = \emptyset.$$

Example: Consider a root system $R = \{\alpha_1, \alpha_2, \alpha_1 + \alpha_2, -\alpha_1, -\alpha_2, -\alpha_1 - \alpha_2\}$ of type A_2. The subsets $S_1 = \{\alpha_1, \alpha_1 + \alpha_2\}, S_2 = \{\alpha_2\}$ form a 2-partition of the set of positive roots.

In Ref. 2, the 2-partitions of the set of positive roots are classified by Papi. The main result in Ref. 2 is the following.

Theorem 3.1. *An ordered subset L of R_+ is equal to $N(w)$ for some $w \in W$ if and only if the following two conditions are verified:*

(1) If $\alpha, \beta \in L, \alpha < \beta$, and $\alpha + \beta \in R$, then $\alpha + \beta \in L$ and $\alpha < \alpha + \beta$.
(2) If $\alpha + \beta \in L, \alpha, \beta \in R$, then α or β (or both) belong to L and one of them precedes $\alpha + \beta$.

Remark 3.2. The two conditions in the previous theorem are equivalent to the condition that the order on L is convex and the complement $L' = R_+ \setminus L$ is closed.

Next we would like to give a simpler proof than the one in Ref. 2. Denote by w_0 the longest element in the Weyl group. The following result gives a complete classification of all of the 2-partitions of R_+.

Theorem 3.2. *The sets S_1 and S_2 form a 2-partition of the set of positive roots R_+ if and only if there is an element $w \in W$ such that $S_1 = N(w), S_2 = N(w_0 w)$.*

Proof. First, if $S_1 = N(w)$, then

$$N(w_0 w) = \{\alpha \in R_+ | w_0 w(\alpha) \in R_-\} = \{\alpha \in R_+ | w(\alpha) \in R_+\} = R_+ \backslash S_1 = S_2,$$

as w_0 turns all positive roots into negative roots and vice versa. So $R_+ = N(w) \cup N(w_0 w)$ is indeed a 2-partition of R_+.

To prove the statement, we use induction on the number of roots in S_1, If either $S_1 = \emptyset$, or $S_2 = \emptyset$, then $w = 1$ or $w = w_0$ and the statement holds.

From now on, we assume that both S_1 and S_2 are non-empty. We claim that

$$S_1 \cap \Pi \neq \emptyset.$$

Otherwise, $\Pi \subset S_2$. As S_2 is closed, we deduced that $S_2 = R_+$ and $S_1 = \emptyset$. If S_1 has only one vector, then it must be a simple roots α and then

$$S_1 = N(s_\alpha), S_2 = N(w_0 s_\alpha).$$

Hence the theorem is correct in this case.

Assume that the statement is true for each partition with $| S_1 |= m$, we show that it is also true for $| S_1 |= m+1$. Choose $\alpha \in S_1 \cap \Pi$ and consider $s_\alpha \in W$. Then $s_\alpha(R_+) = (R_+ \setminus \{\alpha\}) \cup \{-\alpha\}$ by Lemma 2.1.

Now consider the subset $S_1' = R_+ \cap s_\alpha S_1$ and its complement S_2'. Clearly, S_1' contains m roots. We shall show that both of S_1' and S_2' are closed. Because α is simple, it cannot be the sum of two positive roots. Therefore $S_1 \setminus \{\alpha\}$ is a closed set and by Lemma 2.1,

$$s_\alpha(S_1 \setminus \{\alpha\}) = s_\alpha S_1 \setminus \{-\alpha\} = R_+ \cap s_\alpha S_1 = S_1'$$

is also closed.

Next we show that $S_2' = s_\alpha S_2 \cup \{\alpha\}$, is a closed set. To show this we switch it by s_α to obtain $S_2 \cup \{-\alpha\}$. As S_2 itself is closed, the only thing we have to make clear is , for $\beta \in S_2$, $-\alpha + \beta$ must be in S_2 if $-\alpha + \beta$ is a root. In case $-\alpha + \beta \in S_1$, then $(-\alpha + \beta) + \alpha = \beta \in S_1$, a contradiction to our conditions. $\square$

4. 2-partitions of the root system of finite type

In this section, we study the 2-partition of the root system R. Recall that a 2-partition of R is a pair of additionally closed subsets S_1, S_2 satisfying

$$R = S_1 \cup S_2, S_1 \cap S_2 = \emptyset.$$

Example: Consider a root system $R = \{\alpha_1, \alpha_2, \alpha_1 + \alpha_2, -\alpha_1, -\alpha_2, -\alpha_1 - \alpha_2\}$ of type A_2. The subsets $S_1 = \{\alpha_1, \alpha_2, \alpha_1 + \alpha_2, -\alpha_1\}, S_2 = \{-\alpha_2, -\alpha_1 - \alpha_2\}$ is a 2-partition of the root system R.

Lemma 4.1. *Let $R = S_1 \cup S_2$ be a two partition of the root system. Then $R = w(S_1) \cup w(S_2)$ is also a 2-partition of the root system, for any $w \in W$.*

The following lemma will be needed in the classification of the 2-partitions of the root system R.

Lemma 4.2. *Let $\beta \in R_+$ and $i \in supp\beta$. Then there exists a sequence of simple roots $\alpha_{i_1}, \alpha_{i_2}, \cdots, \alpha_{i_m}$ such that β can be written as*

$$\beta = \alpha_{i_1} + \alpha_{i_2} + \cdots + \alpha_{i_m}, \ m = ht\beta$$

such that $i_1 = i$ and $\alpha_{i_1} + \cdots + \alpha_{i_s} \in R_+$ for all $1 \le s \le m$.

Proof. We use induction on $ht\beta$. The case $ht\beta = 1$ is trivial. Assume that the statement is true for $ht\beta \le m - 1$. If $ht\beta = m \ge 2$, there exists $\alpha \in \Pi$ such that $\beta - \alpha \in R_+$. If $i \in supp(\beta - \alpha)$, then the positive root $\beta - \alpha$ can be written as

$$\beta - \alpha = \alpha_{i_1} + \alpha_{i_2} + \cdots + \alpha_{i_{m-1}}$$

with $\alpha_{i_1} = \alpha_i$ and $\alpha_{i_1} + \cdots + \alpha_{i_s} \in R_+$ for all $1 \le s \le m - 1$, by induction hypothesis. Hence

$$\beta = \alpha_{i_1} + \alpha_{i_2} + \cdots + \alpha_{i_{m-1}} + \alpha$$

with $\alpha_{i_1} = \alpha_i$ and $\alpha_{i_1} + \cdots + \alpha_{i_s} \in R_+$ for all $1 \le s \le m$. If $i \notin supp(\beta - \alpha)$, then $\alpha = \alpha_i$ and β can be written as

$$\beta = \sum_{j=1}^{n} a_k \alpha_j$$

with $a_i = 1$. Since $supp\beta$ is connected, there exists $k \in supp\beta$ such that $\langle \alpha_i, \alpha_k \rangle < 0$. Again by the induction hypothesis, the positive root $\beta - \alpha_i$ can be written as

$$\beta - \alpha_i = \alpha_{i_1} + \alpha_{i_2} + \cdots + \alpha_{i_{m-1}}$$

with $i_1 = k$ and $\alpha_{i_1} + \cdots + \alpha_{i_s} \in R_+$ for all $1 \leq s \leq m - 1$, by induction hypothesis. Since

$$\langle \alpha_i, \alpha_{i_1} + \cdots + \alpha_{i_s} \rangle < 0,$$

$\alpha_i + \alpha_{i_1} + \cdots + \alpha_{i_s} \in R_+$ for all $1 \leq s \leq m - 1$ and the expression

$$\beta = \alpha_i + \alpha_{i_1} + \cdots + \alpha_{i_{m-1}}$$

meets our requirement. This completes the process of induction. $\qquad\square$

Remark 4.1. The above lemma does not follows from Lemma A in the section 10.2 of Humphreys' book[8] since i is an arbitrary node in the support of the root β.

Theorem 4.1. *Let $R = S_1 \cup S_2$ be a 2-partition of the root system. Then there is a $w \in W$ and two disconnected sub-diagrams I and J of the Dynkin diagram $D(A)$ such that*

$$w(S_1) = (R_+ \setminus R(I)) \cup R(J), w(S_2) = (R_- \setminus R(J)) \cup R(I).$$

Proof. Choose $w \in W$ such that $\mid w(S_1) \cap R_+ \mid$ is maximal. Then

$$R = w(S_1) \cup w(S_2)$$

is also a 2-partition and $\mid w(S_2) \cap R_- \mid$ is maximal. Without loss of generality, we may assume that $w = 1$ hereafter.

We claim that if $\alpha_i \in S_2 \cap \Pi$, then $-\alpha_i \in S_2 \cap -\Pi$. Indeed, if $-\alpha_i \in S_1$, then

$$\mid s_i(S_1) \cap R_+ \mid = \mid S_1 \cap R_+ \mid + 1,$$

a contradiction!

Similarly, one can show that if $-\alpha_i \in S_1$, then $\alpha_i \in S_1$.

We also claim that if $\beta \in S_2 \cap R_+$, $i \in supp\beta$, then $\alpha_i \in S_2$ and hence $-\alpha_i \in S_2$. By Lemma 4.2, the root β can be written as

$$\beta = \alpha_{i_1} + \alpha_{i_2} + \cdots + \alpha_{i_m}, \ m = ht\beta$$

with $i_1 = i$ and $\alpha_{i_1} + \alpha_{i_2} + \cdots + \alpha_{i_s} \in R_+$ for all $1 \leq s \leq m$. If $\alpha_{i_m} \in S_2$, then $-\alpha_{i_m} \in S_2$. Hence $\alpha_{i_1} + \alpha_{i_2} + \cdots + \alpha_{i_{m-1}} = \beta + (-\alpha_{i_m}) \in S_2$ by the closeness of S_2. If $\alpha_{i_m} \notin S_2$, then the closeness of S_1 forces that $\alpha_{i_1} + \alpha_{i_2} + \cdots + \alpha_{i_{m-1}} \in S_2$. Keep going with $\beta' = \alpha_{i_1} + \alpha_{i_2} + \cdots + \alpha_{i_{m-1}} \in S_2$ and $i \in supp\beta'$, one can show eventually that $\alpha_i \in S_2$.

Let $I = \cup_{\beta \in S_1 \cap R_-} supp\beta$ and let $J = \cup_{\beta \in S_2 \cap R_+} supp\beta$. Then $R(I) \subset S_1$ and $R(J) \subset S_2$. Moreover,

$$S_1 = R(I) \cup (S_1 \cap R_+), \quad S_2 = R(J) \cup (S_2 \cap R_-),$$

and we conclude that

$$S_1 = R(I) \cup (R_+ \setminus R(J)), \quad S_2 = R(J) \cup (R_- \setminus R(I)).$$

If I and J are connected, there exists $i \in I, j \in J$ such that $\langle \alpha_i, \alpha_j \rangle < 0$ and $\alpha_i \in S_1, \alpha_j \in S_2$. Clearly $\alpha_i + \alpha_j \in R_+$. If $\alpha_i + \alpha_j \in S_1$, then

$$\alpha_j = (\alpha_i + \alpha_j) + (-\alpha_i) \in S_1$$

by the closeness of S_1 which is a contradiction. Similarly, one can show that $\alpha_i + \alpha_j \in S_2$ is also impossible.

If I and J are two disconnected sub-diagrams of the Dynkin diagram $D(A)$, it is elementary to check that the pair of subsets $R(I) \cup (R_+ \setminus R(J))$ and $R(J) \cup (R_- \setminus R(I))$ indeed is a 2-partition of the root system R. Hence the classification of the 2-partition of the root system is complete. $\quad\square$

5. 2-partitions of the root system of affine type

In this section, we assume that the root system R is of affine type. We adapt the labeling of the Dynkin diagrams in Ref. 9. For the sake of simplicity, we only consider the affine root systems with Dynkin diagrams in the Table Aff 1 in Ref. 9 and we fix a set of simple roots $\Pi = \{\alpha_0, \alpha_1, \cdots, \alpha_n\}$. Removing the 0-node in the Dynkin diagram of affine type, one gets a Dynkin diagram $D(A^o)$ of finite type and the corresponding root system will be denoted by R^o. Let W^o be the Weyl group associated to R^o. Denote by δ the minimal positive imaginary root. It is well known that the set of imaginary roots is $R^{im} = \{m\delta \mid 0 \neq m \in \mathbb{Z}\}$. Denote by $R_+^{im} = \{m\delta \mid m \in \mathbb{Z}_{>0}\}$ the set of positive imaginary roots and $R_-^{im} = -R_+^{im}$ the set of negative imaginary roots. The following result is well-known (see Ref. 9).

Theorem 5.1. *Let R be a root system with Dynkin diagram in the Table Aff 1 in Ref. 9. Then the set of real roots $R^{re} = R_+^{re} \cup -R_+^{re}$, where $R_+^{re} = \{\alpha + m\delta \mid \alpha \in R^o, m \in \mathbb{Z}_{>0}\} \cup R_+^o$.*

Lemma 5.1. *Let $R = S_1 \cup S_2$ be a 2-partition of R. Then only one of the following four possibilities occurs.*

(1) $R^{im} \subset S_1$ and $R^{im} \cap S_2 = \emptyset$;
(2) $R^{im} \subset S_2$ and $R^{im} \cap S_1 = \emptyset$;

(3) $R_+^{im} \subset S_1$ and $-R_+^{im} \subset S_2$;

(4) $-R_+^{im} \subset S_1$ and $R_+^{im} \subset S_2$.

Proof. If $S_1 \cap R_+^{im} \neq \emptyset$, we consider the minimal positive integer m such that $m\delta \in S_1$. If $m \geq 2$, then $\delta \in S_2$ and hence $R_+^{im} \subset S_2$ by the closeness of S_2. This is a contradiction. Hence $m = 1$ and $R_+^{im} \subset S_1$. Similarly, $-R_+^{im} \subset S_1$ if $S_1 \cap -R_+^{im} \neq \emptyset$. $\square$

If the subsets S_1 and S_2 form a 2-partition of R, then the subsets $S_1 \cap R^o$ and $S_2 \cap R^o$ form a 2-partition of the finite type root system R^o. Hence by Theorem 4.1, there is an element $w \in W^o$ and two disconnected sub-diagrams I and J of $D(A^o)$ such that

$$S_1 \cap R^o = w(R(I) \cup (R_+^o \setminus R(J))), S_2 \cap R^o = w(R(J) \cup (R_-^o \setminus R(J))).$$

Since both R_+^{im} and R_-^{im} are W^o-invariant, we may assume from now on that $w = 1$. Also, we may only consider the cases (1) and (3) in the previous lemma. The cases (2) and (4) in the previous lemma can be treated symmetrically to the cases (1) and (3) respectively.

If $R^{im} \subset S_1$, $R^{im} \cap S_2 = \emptyset$, then $R^{im} \cup \{\alpha + m\delta \mid m \in \mathbb{Z}, \alpha \in R(I) \cup (R_+^o \setminus R(J))\} \subset S_1$. If $J \neq \emptyset$, there is a simple root α_i such that both α_i and $-\alpha_i \in S_2$. Since $\delta = (\alpha_i + \delta) + (-\alpha_i) \in S_1$, we have $\alpha_i + \delta \in S_1$ which forces $\alpha_i = (\alpha_i + \delta) + (-\delta) \in S_1$, this is a contradiction. Hence, $J = \emptyset$. Clearly,

$$R^{im} \cup \{\alpha + m\delta \mid \alpha \in R(I) \cup R_+^o, m \in \mathbb{Z}\} \subset S_1$$

and

$$\{\beta + m\delta \mid \beta \in R_-^o \setminus R(I), m \in \mathbb{Z}\} \subset S_2.$$

We conclude that the equalities hold.

Now, we assume that $R_+^{im} \subset S_1$ and $-R_+^{im} \subset S_2$. Since

$$S_1 \cap R^o = R(I) \cup (R_+^o \setminus R(J)), S_2 \cap R^o = R(J) \cup (R_-^o \setminus R(J)),$$

we get

$$S_1 \supset R(I) \cup (R_+^o \setminus R(J)) + \mathbb{Z}_+\delta$$

and

$$S_2 \supset R(J) \cup (R_-^o \setminus R(J)) - \mathbb{Z}_+\delta.$$

For any $\alpha \in R(I)$, both $\alpha + m\delta$ and $-\alpha + m\delta$ are in S_1 by the closeness of S_1, for any $m \in \mathbb{Z}_{>0}$. Since $-m\delta = -\alpha + (\alpha - m\delta) \in S_2$, the real root

$\alpha - m\delta \in S_2$ for all $\alpha \in R(I)$. Similarly, we have $\beta + m\delta \in S_1$ for all $\beta \in R(J)$ and $m \in \mathbb{Z}_{>0}$. Therefore, $R(J) + \mathbb{Z}_{>0}\delta \subset S_1$ and $R(I) - \mathbb{Z}_{>0}\delta \subset S_2$. Hence we have

$$S_1 \supset (R(I) \cup (R_+^o \setminus R(J)) + \mathbb{Z}_+\delta) \cup (R(J) + \mathbb{Z}_{>0}\delta)$$

and

$$S_2 \supset (R(J) \cup (R_-^o \setminus R(J)) - \mathbb{Z}_+\delta) \cup (R(I) - \mathbb{Z}_{>0}\delta).$$

The other undetermined roots form a subset of R

$$(R_+^o \setminus (R(I) \cup R(J)) - \mathbb{Z}_{>0}\delta) \cup (R_-^o \setminus (R(J) \cup R(J)) + \mathbb{Z}_{>0}\delta).$$

To completely classify 2-partitions in this case, we need several definitions as follows.

Definition 5.1.

(1) A real root α is called **bad** in S_i (with respect to the 2-partition (S_1, S_2)) if $\alpha + \mathbb{Z}\delta \subset S_i$.
(2) A real root α is called **good** if it is not bad and there exists an $m \in \mathbb{Z}$ such that $\alpha - m\delta \notin S_1$.
(3) A real root α is called **very good** if it is good and $\alpha = \sum_{i=0}^n m_i \alpha_i$ with $m_0 < 0$ and $-\alpha \in S_1$.

One can see that $\alpha \in R(I) \cup R(J)$ is good. It is also not difficult to check the following lemma.

Lemma 5.2.

(1) For two bad roots α, $\beta \in S_i$, if $\alpha + \beta \in R^{re}$, then $\alpha + \beta$ is a bad root in S_i.
(2) For a bad root $\alpha \in S_i$ and any root $\beta \in S_i \cup (-1)^i R_-^o$, if $\alpha + \beta \in R^{re}$, then $\alpha + \beta$ is a bad root in S_i.
(3) For two good roots α, β, if $\alpha + \beta \in R^{re}$, then $\alpha + \beta$ is a good root.

If α is bad is S_1, then $-\alpha + m\delta \notin S_1$. Otherwise $\mathbb{Z}\delta \subset S_1$ by the closeness of S_1. Hence $-\alpha + \mathbb{Z}\delta \subset S_2$. Similarly, if α is bad in S_2, then $-\alpha$ is bad in S_1.

We denote by C^b the subset of R_+^o consisting of all bad roots in S_1. Let Π^b be the set of all bad simple roots in R^o which is indexed by I^b, i.e. $\Pi^b = C^b \cap \Pi = \{\alpha_i \mid i \in I^b\}$. We claim that

$$C^b = \{\alpha \in R_+^o \mid supp\alpha \cap I^b \neq \emptyset\}.$$

Indeed, for $\alpha \in R_+^o$ with $supp\alpha \cap I^b \neq \emptyset$, there is an $i \in supp\alpha \cap I^b$. It follows immediately from Lemma 4.2 and Lemma 5.2(2) that α is bad in S_1. Conversely, if $\alpha \in C^b$, there exists a sequence of simple roots $\alpha_{i_1}, \alpha_{i_2}, \cdots, \alpha_{i_m}$ in R^0 such that $\alpha = \alpha_{i_1} + \alpha_{i_2} + \cdots + \alpha_{i_m}$ and $\alpha_{i_1} + \cdots + \alpha_{i_s} \in R_+^o$ for all $1 \leq s \leq m$. If α_{i_s} is good for all $1 \leq s \leq m$, then α is good by Lemma 5.2(3) which is a contradiction. Hence α_{i_s} is bad for some $1 \leq s \leq m$. Moreover, one has

$$\alpha_{i_s} \in (R_+^o \setminus R(J)) \subset S_1.$$

It follows that $i_s \in supp\alpha \cap I^b \neq \emptyset$. Hence the equality holds.

Once I^b is determined, so is C^b and from the above argument, we have

$$S_1 \supset C^b + \mathbb{Z}\delta \quad \text{and} \quad S_2 \supset -C^b + \mathbb{Z}\delta.$$

The index set I^b of bad simple roots in R_+^0 can be, in fact, arbitrarily chosen. Removing all the nodes in I^b from the Dynkin diagram $D(A^o)$ of finite type, one obtains a disconnected union of connected sub-diagrams of $D(A^o)$, written as

$$I_1 \cup I_2 \cup \cdots \cup I_m.$$

Note that each I_s is a Dynkin diagram of finite type and $I \cup J \subset I_1 \cup I_2 \cup \cdots \cup I_m$. Moreover, it is known that

$$R(I_1)_+ \cup R(I_2)_+ \cup \cdots \cup R(I_m)_+ \cup C^b = R_+^o$$

is a disjoint union. Hence the only thing we need to do is to determine what the set $R(I_1) \cup R(I_2) \cup \cdots \cup R(I_m) + \mathbb{Z}\delta$ of all good roots in R will be like.

It is easy to see, from the Definition 5.1(3), that there are only finitely many very good roots in S_1. We choose $m > 0$ maximal such that $\alpha = \alpha^o - m\delta \in S_1$ is very good where $\alpha^o \in R^o$.

Lemma 5.3. $\{\beta \in s_\alpha S_1 \mid \beta$ is very good with respect to $(s_\alpha S_1, s_\alpha S_2)\}$ is a proper subset of $\{\beta \in S_1 \mid \beta$ is very good with respect to $(S_1, S_2)\}$ where α is fixed as above.

Proof. For any $\gamma \in S_1$ such that $s_\alpha(\gamma)$ is very good with respect to $(s_\alpha S_1, s_\alpha S_2)$, it can be easily seen that γ is good and $-\gamma \in S_2$. If $\langle \gamma, \alpha \rangle < 0$, then $\gamma + \alpha \in R$ and moreover, $\gamma + \alpha \in S_1$. Since $-\alpha \in S_2$, we have $-\gamma - \alpha \in S_2$. One can show consecutively that for $1 \leq i \leq -\langle \gamma, \alpha \rangle$, $\gamma + i\alpha \in S_1$ and $-\gamma - i\alpha \in S_1$. In particular,

$$s_\alpha(\gamma) = \gamma - \langle \gamma, \alpha \rangle \alpha \in S_1 \quad \text{and} \quad -s_\alpha(\gamma) \in S_2.$$

Hence $s_\alpha(\gamma) \in S_1$ is very good with respect to 2-partition (S_1, S_2).

If $\langle \gamma, \alpha \rangle \geq 0$, we write γ as

$$\gamma = \sum_{0 \leq i \leq n} n_i \alpha_i$$

and then

$$n_0 + \langle \gamma, \alpha \rangle m < 0.$$

It follows that $n_0 < 0$ and moreover, $\gamma \in S_1$ is very good with respect to (S_1, S_2). From the choice of m, one obtains that $\langle \gamma, \alpha \rangle = 0$. Hence $s_\alpha(\gamma) = \gamma$ is very good with respect to (S_1, S_2).

The inequality holds since obviously α is contained in the second set but not in the first one. $\qquad \square$

Note that the action of the elements in W^o does not produce very good roots. Applying the simple reflections determined by very good roots α as above, we may assume from now on that there is no very good roots in S_1.

If $\alpha \in R_+^o \setminus (R(I) \cup R(J))$ is good in S_1, then $\alpha - m\delta \in S_2$ for some integer m which is necessarily positive. We may assume that m is minimal with respect to the above requirement. If $m \geq 3$, then $\alpha - \delta, \alpha - 2\delta \in S_1$ and $-\alpha + \delta, -\alpha + 2\delta \in S_1$ since there is no very good roots in S_1. But then $-\delta = (-\alpha + \delta) + (\alpha - 2\delta) \in S_1$. This is a contradiction. Hence, $m = 1$ or 2.

If $m = 1$, then $\alpha - \delta \in S_2$ and moreover $\alpha - \mathbb{Z}_{>0}\delta \subset S_2$. Clearly, $-\alpha + \delta + \mathbb{Z}_{>0}\delta \subset S_1$, otherwise $\delta \in S_2$. But $-\alpha + \delta$ can be in S_1 or S_2. α is said to be 0-**good** (resp. (−1)-**good**) if $-\alpha + \delta \in S_1$ (resp. $-\alpha + \delta \in S_2$). Note that $-\alpha - \mathbb{Z}_+\delta \subset S_2$ if α is 0-good; $-\alpha + \delta - \mathbb{Z}_+\delta \subset S_2$ if α is (−1)-good.

If $m = 2$, then $\alpha - \delta \in S_1$ which forces $-\alpha + \delta \in S_1$. Hence,

$$(\alpha - \delta + \mathbb{Z}_+\delta) \cup (-\alpha + \mathbb{Z}_{>0}\delta) \subset S_1$$

and

$$(-\alpha - \mathbb{Z}_+\delta) \cup (\alpha - \delta - \mathbb{Z}_{>0}\delta) \subset S_2.$$

We say α is 1-**good** in this case. Also we define $\alpha \in R(I)$ (resp. $\alpha \in R(J)$) to be I-**good** (resp. J-**good**). Note that 0-good, ± 1-good and I-good roots are all contained in S_1 while J-good roots are in S_2.

Lemma 5.4. *For two roots* $\alpha, \beta \in R^o$,

(1) if α is 1-good and β is ± 1-good or J-good, then $\alpha + \beta$ is not a root;

(2) if α is 1-good, β is 0-good or I-good and if $\alpha + \beta$ is a root, then $\alpha + \beta$ is 1-good;

(3) if α is 0-good and β is (-1)-good, then $\alpha + \beta$ is not a root;

(4) if α is 0-good, β is I-good or J-good and if $\alpha + \beta$ is a root, then $\alpha + \beta$ is 0-good;

(5) if both α and β are 0-good and if $\alpha + \beta$ is a root, then $\alpha + \beta$ is either 0-good or ± 1-good;

(6) if α is (-1)-good and β is (-1)-good or I-good, then $\alpha + \beta$ is not a root;

(7) if α is (-1)-good, β is J-good and if $\alpha + \beta$ is a root, then $\alpha + \beta$ is (-1)-good;

(8) if both α and β are I-good (resp. J-good) and if $\alpha + \beta$ is a root, then $\alpha + \beta$ is I-good (resp. J-good);

(9) if α is I-good and β is J-good, then $\alpha + \beta$ is not a root.

Proof. We know already from Lemma 5.2 that the sum of two good roots in R^o is also good if it is a root.

(1) If α is 1-good and so is β, then $\alpha - \delta, \beta - \delta \in S_1$. Assume that $\alpha + \beta$ is a root in R^o, $\alpha + \beta - 2\delta$ is then a root in R. Hence $\alpha + \beta - 2\delta \in S_1$ by the closeness of S_1 which contradicts with $m = 1, 2$. If β is (-1)-good, then $\alpha - \delta \in S_1$ and $\beta \in S_1$. It follows that $\alpha + \beta - \delta \in S_1$ if $\alpha + \beta$ is a root. Hence $\alpha + \beta \in S_1$ must be 1-good and $-\alpha - \beta + \delta \in S_1$. Since $\alpha \in S_1$, we have $-\beta + \delta \in S_1$ which is a contradiction. We assume that β is J-good. If $\alpha + \beta - \delta \in S_1$, one obtains that $\beta \in S_1$ since $-\alpha + \delta \in S_1$ which is absurd. Hence $\alpha + \beta - \delta \in S_2$. Since $-\beta \in R(J) \subset S_2$, it forces $\alpha - \delta \in S_2$ which is a contradiction. Hence $\alpha + \beta$ is not a root.

(2) Assume that α is 1-good, then $\alpha - \delta \in S_1$. If β is 0-good or I-good, $\beta \in S_1$. Hence $\alpha + \beta - \delta \in S_1$ once $\alpha + \beta$ is a root. It implies that $\alpha + \beta$ is 1-good.

(3) $-\beta + \delta \in S_2$ since β is (-1)-good. Also we have $-\alpha \in S_2$ since α is 0-good. Hence $-\alpha - \beta + \delta \in S_2$ if $\alpha + \beta$ is a root. It follows that $\alpha + \beta - \delta \in S_2$ since otherwise $\alpha + \beta - \delta \in S_1$ is very good. By the closeness of S_2, we have $\alpha \in S_2$, a contradiction. Hence $\alpha + \beta$ is not a root.

(4) We assume that α is 0-good. If β is I-good, $-\beta \in R(I) \subset S_1$. Clearly, $\alpha + \beta - \delta \in S_2$ if $\alpha + \beta$ is a root. Otherwise if $\alpha + \beta - \delta \in S_1$, then $\alpha - \delta \in S_1$ which contradicts that α is 0-good. Hence $\alpha + \beta$ is 0-good or (-1)-good. Since $-\alpha + \delta \in S_1$, one has $-\alpha - \beta + \delta \in S_1$ which implies $\alpha + \beta$ is 0-good. If β is J-good, then $\beta \in S_2$. Since $\alpha - \delta \in S_2$, $\alpha + \beta - \delta \in S_2$ if $\alpha + \beta$ is a root. Hence $\alpha + \beta$ is not 1-good. If $-\alpha - \beta + \delta \in S_2$, then $-\alpha + \delta \in S_2$

which is absurd. Hence $-\alpha - \beta + \delta \in S_1$ and it implies that $\alpha + \beta$ is 0-good.

(5) It is clear that $\alpha + \beta$ cannot be I-good or J-good.

(6) Assume that α is (-1)-good. We have $-\beta + \delta \in S_2$ for a (-1)-good root β. If $\alpha + \beta$ is a root and furthermore, $\alpha + \beta - \delta \in S_2$, then $\alpha \in S_2$ which is absurd. Hence $\alpha + \beta - \delta \in S_1$ and moreover, $\alpha + \beta$ is 1-good. We have $-\alpha - \beta + \delta \in S_1$. Since $\alpha \in S_1$, $-\beta + \delta \in S_1$ which contradicts that β is (-1)-good. Hence $\alpha + \beta$ is not a root. If β is I-good, one gets $-\beta \in S_1$. Since $\alpha - \delta \notin S_1$, it forces $\alpha + \beta - \delta \notin S_1$. Hence $\alpha + \beta - \delta \in S_2$ if $\alpha + \beta$ is a root. Clearly we have $-\alpha + \delta \in S_2$ and it follows that $\beta \in S_2$ which is a contradiction. Thus $\alpha + \beta$ is not a root.

(7) Clearly, $\pm(\alpha - \delta) \in S_2$, $\pm\beta \in S_2$ for a (-1)-good root α and a J-good root β. Hence $\pm(\alpha + \beta - \delta) \in S_2$ and $\alpha + \beta$ is (-1)-good if it is a root.

(8) and (9) are obvious. □

Several examples for Lemma 5.4(5) will be given below while we mention here that all possibilities in (2), (4), (7), (8) could happen.

Example: Consider a root system R of type $\widetilde{A}_2$. Then $R^o = \{\pm\alpha_1, \pm\alpha_2, \pm(\alpha_1 + \alpha_2)\}$ and assume α_i is 0-good for $i = 1, 2$. The following S_1 and S_2 form a 2-partition of R.

(1) $\alpha_1 + \alpha_2$ is 1-good.

$$S_1 = \mathbb{Z}_{>0}\delta \cup (\alpha_1 + \mathbb{Z}_+\delta) \cup (-\alpha_1 + \mathbb{Z}_{>0}\delta) \cup (\alpha_2 + \mathbb{Z}_+\delta)$$
$$\cup (-\alpha_2 + \mathbb{Z}_{>0}\delta) \cup (\alpha_1 + \alpha_2 - \delta + \mathbb{Z}_+\delta) \cup (-\alpha_1 - \alpha_2 + \mathbb{Z}_{>0}\delta),$$

$$S_2 = (-\mathbb{Z}_{>0}\delta) \cup (\alpha_1 - \mathbb{Z}_{>0}\delta) \cup (-\alpha_1 - \mathbb{Z}_+\delta) \cup (\alpha_2 - \mathbb{Z}_{>0}\delta)$$
$$\cup (-\alpha_2 - \mathbb{Z}_+\delta) \cup (\alpha_1 + \alpha_2 - 2\delta - \mathbb{Z}_+\delta) \cup (-\alpha_1 - \alpha_2 - \mathbb{Z}_+\delta).$$

(2) $\alpha_1 + \alpha_2$ is 0-good.

$$S_1 = \mathbb{Z}_{>0}\delta \cup (\alpha_1 + \mathbb{Z}_+\delta) \cup (-\alpha_1 + \mathbb{Z}_{>0}\delta) \cup (\alpha_2 + \mathbb{Z}_+\delta)$$
$$\cup (-\alpha_2 + \mathbb{Z}_{>0}\delta) \cup (\alpha_1 + \alpha_2 + \mathbb{Z}_+\delta) \cup (-\alpha_1 - \alpha_2 + \mathbb{Z}_{>0}\delta),$$

$$S_2 = (-\mathbb{Z}_{>0}\delta) \cup (\alpha_1 - \mathbb{Z}_{>0}\delta) \cup (-\alpha_1 - \mathbb{Z}_+\delta) \cup (\alpha_2 - \mathbb{Z}_{>0}\delta)$$
$$\cup (-\alpha_2 - \mathbb{Z}_+\delta) \cup (\alpha_1 + \alpha_2 - \mathbb{Z}_{>0}\delta) \cup (-\alpha_1 - \alpha_2 - \mathbb{Z}_+\delta).$$

(3) $\alpha_1 + \alpha_2$ is (-1)-good.

$$S_1 = \mathbb{Z}_{>0}\delta \cup (\alpha_1 + \mathbb{Z}_+\delta) \cup (-\alpha_1 + \mathbb{Z}_{>0}\delta) \cup (\alpha_2 + \mathbb{Z}_+\delta)$$
$$\cup (-\alpha_2 + \mathbb{Z}_{>0}\delta) \cup (\alpha_1 + \alpha_2 + \mathbb{Z}_+\delta) \cup (-\alpha_1 - \alpha_2 + 2\delta + \mathbb{Z}_+\delta),$$

$$S_2 = (-\mathbb{Z}_{>0}\delta) \cup (\alpha_1 - \mathbb{Z}_{>0}\delta) \cup (-\alpha_1 - \mathbb{Z}_{+}\delta) \cup (\alpha_2 - \mathbb{Z}_{>0}\delta)$$
$$\cup (-\alpha_2 - \mathbb{Z}_{+}\delta) \cup (\alpha_1 + \alpha_2 - \mathbb{Z}_{>0}\delta) \cup (-\alpha_1 - \alpha_2 + \delta - \mathbb{Z}_{+}\delta).$$

For simplicity, we list all the possibilities for the sum of two good roots in $R_+^o \cup R(I) \cup R(J)$ in the following table.

Table 1.   Sum of 2 good roots

	1	0	−1	I	J
1	X	1	X	1	X
0	1	±1, 0	X	0	0
−1	X	X	X	X	−1
I	1	0	X	I	X
J	X	0	−1	X	J

Note: "X" means that the sum is not a root.

It is known before that the only undetermined elements in R are those in

$$\pm R_+^o \setminus (R(I) \cup R(J) \cup C^b) \mp \mathbb{Z}_{>0}\delta$$

and since $\alpha \in R_+^o \setminus (R(I) \cup R(J) \cup C^b)$ is either ± 1-good or 0-good, it turns out that we only need to make a decision for each α. Denote by Ψ the map

$$\Psi : R_+^o \setminus (R(I) \cup R(J) \cup C^b) \longrightarrow \{1, 0, -1\}$$

which takes α to to $i \in \{1, 0, -1\}$ if α is i-good. One can see from Lemma 5.4 that Table 1 holds. We conclude that the converse is also true. To be precise, let Φ be a map

$$\Phi : R_+^o \setminus (R(I) \cup R(J) \cup C^b) \longrightarrow \{1, 0, -1\}.$$

Φ is said to be an **admissible choice** if Table 1 holds when any $\alpha \in R_+^o \setminus (R(I) \cup R(J) \cup C^b)$ is $\Phi(\alpha)$-good. It is not difficult to check that if Φ is an admissible choice (I, J and C^b are fixed), then the associated pair (S_1, S_2) is indeed a 2-partition.

Now, we have the following theorem to classify 2-partitions of affine root system.

Theorem 5.2. *Let R be a root system with Dynkin diagram in the Table Aff 1 in Ref. 9. The subsets S_1 and S_2 form a 2-partition of R if and only if there is an element $w \in W$, two disconnected sub-diagrams I and J of the Dynkin diagram $D(A^o)$, a set of bad simple roots $\Pi^b = \{\alpha_i \in \Pi^o \mid i \in$*

$I^b\} \subset (R^o_+ \setminus (R(I) \cup R(J)))$ *and an admissible choice* Φ *such that one of the following conditions is satisfied up to a permutation of* S_1 *and* S_2.

(1) $wS_1 = R^{im} \cup (R(I) \cup R^o_+ + \mathbb{Z}\delta); \quad wS_2 = (R^o_- \setminus R(I)) + \mathbb{Z}\delta.$

(2)

$$wS_1 = R^{im}_+ \cup (R(I) + \mathbb{Z}_+\delta) \cup (R(J) + \mathbb{Z}_{>0}\delta) \cup (C^b + \mathbb{Z}\delta)$$
$$\cup (\Phi^{-1}(1) - \delta + \mathbb{Z}_+\delta) \cup (-\Phi^{-1}(1) + \mathbb{Z}_{>0}\delta)$$
$$\cup (\Phi^{-1}(0) + \mathbb{Z}_+\delta) \cup (-\Phi^{-1}(0) + \mathbb{Z}_{>0}\delta)$$
$$\cup (\Phi^{-1}(-1) + \mathbb{Z}_+\delta) \cup (-\Phi^{-1}(-1) + \delta + \mathbb{Z}_{>0}\delta);$$

$$wS_2 = R^{im}_- \cup (R(J) - \mathbb{Z}_+\delta) \cup (R(I) - \mathbb{Z}_{>0}\delta) \cup (-C^b + \mathbb{Z}\delta)$$
$$\cup (\Phi^{-1}(1) - \delta - \mathbb{Z}_{>0}\delta) \cup (-\Phi^{-1}(1) - \mathbb{Z}_+\delta)$$
$$\cup (\Phi^{-1}(0) - \mathbb{Z}_{>0}\delta) \cup (-\Phi^{-1}(0) - \mathbb{Z}_+\delta)$$
$$\cup (\Phi^{-1}(-1) - \mathbb{Z}_{>0}\delta) \cup (-\Phi^{-1}(-1) + \delta - \mathbb{Z}_+\delta).$$

where $C^b = \{\alpha \in R^o_+ \mid supp\alpha \cap I^b \neq \emptyset\}$ *and all* $\cup$*'s are disjoint unions.*

The above result on classification of 2-partitions is not satisfactory since it still remains as a problem to find all the admissible choices. In order to solve this completely, we will give an algorithm to determine all of the admissible choices at the end of this article.

Recall that it has been pointed out previously that

$$(R(I)_+ \cup R(J)_+) \subset (R^o_+ \setminus C^b) = \cup_{1 \leq s \leq m} R(I_s)_+.$$

We denote by θ_s the highest root in $R(I_s)_+$. If there is a (-1)-good root $\alpha \in R(I_s)_+$, it can be seen from Lemma 5.4 that θ_s is (-1)-good. Similarly, θ_s is 1-good if there is a 1-good root in $R(I_s)_+$. Hence (-1)-good and 1-good roots cannot occur in $R(I_s)_+$ simultaneously. Assume that $\alpha \in R(I_s)_+$ is (-1)-good, it follows from Lemma 5.4 that α is of the form

$$\theta_s \overset{\cdot}{-} \sum \beta$$

where all β's in the sum are in $R(J)_+ \cap R(I_s)_+$. We claim that the converse is also true. Indeed, if there exist (-1)-good elements in $R(I_s)_+$ and $\alpha = \theta_s - \sum \beta \in R(I_s)_+$ is a root for some $\beta \in R(J)_+ \cap R(I_s)_+$, clearly it cannot be I-good or J-good. If α is 0-good, consecutively adding $\beta \in R(J)_+$ to it, one obtains θ_s which is, by Lemma 5.4, 0-good. This contradicts the argument above and hence α must be (-1)-good. Denote by $C^i(I_s)$ the set of all i-good roots in $R(I_s)_+$ for $i \in \{1, 0, -1\}$. We have proved

$$C^{-1}(I_s) = \{\alpha = \theta_s - \sum \beta \in R(I_s)_+ \mid \beta \in R(J)_+ \cap R(I_s)_+\}$$

if there exist (-1)-good roots in $R(I_s)_+$. One can see that

$$C^0(I_s) = R(I_s)_+ \setminus (R(I) \cup R(J) \cup C^{-1}(I_s)) \text{ and } C^1(I_s) = \emptyset$$

if that is the case.

From now on we assume that there is no (-1)-good root in $R(I_s)_+$, i.e. $C^{-1}(I_s) = \emptyset$.

Definition 5.2. For $\beta \in R(I_s)_+$, a sequence of roots $\mathscr{S} = (\beta_1, \beta_2, \cdots, \beta_l)$ is called a β-**sequence** if $\beta_i \in R(I_s)_+ \cup R(I) \cup R(J)$, $\beta_1 + \beta_2 + \cdots + \beta_i \in R(I_s)_+ \cup R(I) \cup R(J)$ for $1 \leq i \leq l$ and $\beta_1 + \beta_2 + \cdots + \beta_l = \beta$.

We denote by $Seq\beta$ the set of all β-sequences for a given $\beta \in R(I_s)_+$. For any $\mathscr{S} = (\beta_1, \beta_2, \cdots, \beta_l) \in seq\beta$, let β_t be the rightmost element in the sequence such that $\beta_t \in R(J)$. Define a subset $C^0(I_s, 0)$ of $R(I_s)_+$ to be consisting of all elements $\alpha \notin R(I) \cup R(J)$ such that one of the following conditions is satisfied.

(1) α occurs in the subsequence $(\beta_1, \beta_2, \cdots, \beta_{t-1})$ of $\mathscr{S} \in seq\beta$ for some $\beta \in R(I_s)_+$.
(2) There exists an α-sequence $\mathscr{S}$ which ends with an element in $R(J)$.
(3) α occurs in a β-sequence for some $\beta \in R(I_s)_+$ such that any element ahead of α in the sequence is in $R(J)$.

It is easy to see that $C^0(I_s, 0) \subset C^0(I_s)$. Given a set $A \subset R(I_s)_+ \setminus (R(I) \cup R(J))$, we denote by $\overline{A}$ the minimal set $S \subset R(I_s)_+ \setminus (R(I) \cup R(J))$ containing A such that

(1) $\alpha \in S$ if there exists a β-sequence $\mathscr{S}$ for some $\beta \in S$ such that α occurs in $\mathscr{S}$.
(2) $\alpha \in S$ if there is only one element belong to S in some $\mathscr{S} \in seq\alpha$ and the others in $\mathscr{S}$ are in $R(I) \cup R(J)$.

Set $C^0(I_s, 1) := \overline{C^0(I_s, 0)}$. Clearly $C^0(I_s, 1) \subset C^0(I_s)$. let $\prec$ be any total order on $R(I_s)_+$ such that

$$\alpha \prec \beta \text{ if } ht\alpha < ht\beta \text{ for } \alpha, \beta \in R(I_s)_+.$$

Set $C^1(I_s, 1) := \emptyset$. Choose

$$\beta \in R(I_s)_+ \setminus (R(I) \cup R(J) \cup C^0(I_s, i) \cup C^1(I_s, i))$$

such that β is minimal with respect to $\prec$. The **detecting set** of β, denoted by $D(\beta)$, is defined to be the minimal set $S \subset R(I_s)_+ \setminus (R(I) \cup R(J))$ containing β such that $\alpha \in S$ if there is a $\gamma \in S$ occurring in some $\mathscr{S} \in seq\alpha$.

If there is an α-sequence $\mathscr{S}$ for some $\alpha \in R(I_s)_+$ such that elements in $D(\beta)$ occur more than two times in $\mathscr{S}$, we say $D(\beta)$ is incompatible. Otherwise, $D(\beta)$ is said to be compatible. It is clear that β must be 0-good if $D(\beta)$ is incompatible. When the detecting set $D(\beta)$ is compatible, one can decide arbitrarily whether β is 0-good or 1-good.

If $D(\beta)$ is incompatible or β is chosen to be 0-good, then we define

$$C^0(I_s, i+1) := \overline{C^0(I_s,i) \cup \{\beta\}} \quad \text{and} \quad C^1(I_s, i+1) := C^1(I_s, i).$$

If we choose β to be 1-good when $D(\beta)$ is compatible, let $T(\beta)$ be the subset of $R(I_s)_+ \setminus (R(I) \cup R(J) \cup D(\beta))$ consisting of all elements α such that α occurs in a γ-sequence $\mathscr{S}$, for some $\gamma \in R(I)_+$, which also contains an element in $D(\beta)$. Define in this case

$$C^0(I_s, i+1) := \overline{C^0(I_s,i) \cup T(\beta)} \quad \text{and} \quad C^1(I_s, i+1) := C^1(I_s,i) \cup D(\beta).$$

It is not difficult to see that

$$C^0(I_s, i) \subset C^0(I_s) \quad \text{and} \quad C^1(I_s, i) \subset C^1(I_s)$$

for all $i \in \mathbb{Z}_{>0}$. Since obviously $C^0(I_s, i) \cup C^1(I_s, i)$ is a proper subset of $C^0(I_s, i+1) \cup C^1(I_s, i+1)$, our algorithm stops when

$$C^0(I_s, i) \cup C^1(I_s, i) = R(I_s)_+ \setminus (R(I) \cup R(J)).$$

It follows immediately that

$$C^0(I_s) = C^0(I_s, i), \quad C^1(I_s) = C^1(I_s, i)$$

for this i. We denote by C^j the set of all j-good roots in R_+^0 for $j \in \{1, 0, -1\}$. Hence

$$C^j = \cup_{1 \le s \le m} C^j(I_s) = \Phi^{-1}(j)$$

and the admissible choice Φ is defined finally.

Remark 5.1.

(1) Fixing I, J and C^b, any admissible choice Φ can be actually produced by this algorithm.

(2) Given $\beta \in R(I_s)_+$, a β-sequence $\mathscr{S} = (\beta_1, \beta_2, \cdots, \beta_s)$ is called a strong β-sequence if $\beta_i \in R(I_s)_+$ for all $1 \le i \le l$. Denote by $seq_+\beta$ the set of all strong β-sequences. Replacing all the seq by seq_+ in the previous algorithm, it is not difficult to check that Table 1 still holds. Hence one obtains a modified algorithm which is more convenient since $seq_+\beta$ is a finite set.

Acknowledgments

Partially supported by Chinese National Natural Science Foundation project No. 10631010.

References

1. K. Ito, *Comm. Algebra* **29**, 5605 (2001).
2. P. Papi, A characterization of a special ordering in a root system, in *Proceedings of the A.M.S.*, 1994.
3. P. Papi, *J. Algebra* **172**, 613 (1995).
4. P. Papi, *J. Algebra* **186**, 72 (1996).
5. A. Pilkington, *Comm. Algebra* **34**, 3183 (2006).
6. H. P. Jakobsen and V. G. Kac, A new class of unitarizable highest weight representations of infinite-dimensional lie algebras, in *Nonlinear equations in classical and quantum field theory (Meudon/Paris, 1983/1984)* (Springer, Berlin, 1985).
7. H. P. Jakobsen and V. G. Kac, *J. Funct. Anal.* **82**, 69 (1989).
8. J. Humphreys, *Introduction to Lie Algebras and Representation Theory* (Springer, New York, 1972).
9. V. G. Kac, *Infinite dimensional Lie algebras*, 3rd edn. (Cambridge University Press, 1994).

A SURVEY ON WEAK HOPF ALGEBRAS

FANG LI

Department of Mathematics, Zhejiang University,
Hangzhou, Zhejiang 310027, China
E-mail: fangli@zju.edu.cn

QINXIU SUN*

Department of Mathematics, Zhejiang University,
Hangzhou, Zhejiang 310027, China
E-mail: qinxiusun@126.com

The theory of weak Hopf algebras is a development of the theory of Hopf algebras, which is important in the theory of quantum groups and the related theory of mathematical physics. In this paper, we give a survey of this theory and its applications.

Firstly, we review some concepts related to weak Hopf algebras and their relationship and moreover, some algebraical structures. Secondly, we collect the generalizations of quantum double, i.e. quantum quasi-double and quantum G-double, constructed from some certain weak Hopf algebras so as to obtain a class of singular solutions of the quantum Yang-Baxter equation (QYBE). Their properties are also introduced in this part. Thirdly, as the important examples of weak Hopf algebras, some weaken quantized enveloping algebra are given which are associated with some generalizations of Lie algebra, such as generalized Kac-Moody algebra, Borcherds superalgebras. Moreover, their algebraic structures are discussed such that one can understand them clearly. Finally, some generalized notions of tensor category are introduced including pre-tensor category and weak tensor category and their braidings. They can be realized respectively from the categories of representations of the related classes of weak Hopf algebras and their quantum quasi-doubles. As the categorical version of a (singular) solutions of the QYBE, a general Yang-Baxter operator is studied. In summary, we list the diagrams of relations among these kinds of categories.

Keywords: Weak Hopf algebra; quantum double; quantum Yang-Baxter equation; singular solution; weak quantum algebra; Kac-Moody algebra; pre-tensor category.

*Corresponding author.

Introduction and Algebraic Structures

Because of the important role of Hopf algebra in the theory of quantum group and related mathematical physics, along with the deepening of researches, the meaning of some weaker concepts of Hopf algebra is understood and is paid close attention more and more. A well known concept is the so-called almost bialgebra (i.e. for a field k and a k-linear space H, if (H, μ, η) is a k-algebra and (H, Δ, ε) is a k-coalgebra with $\Delta(xy) = \Delta(x)\Delta(y)$ for $x, y \in H$, then we call H a k-almost bialgebra [9, 11, 6] or a weak bialgebra [24], which is introduced by Hayashi [6] in 1991 so as to construct a face algebra. This construction is developed and applied in a series of works. In [24] and [2], an axiomatic method is used to classify the class of almost bialgebras where an almost bialgebra is called a weak bialgebra. Assume that H is an almost bialgebra, $S : H \longrightarrow H$ a k-linear map satisfying

$$\sum_a a'S(a'') = \sum_1 \varepsilon(1'a)1''$$

and

$$\sum_a S(a')a'' = \sum_1 1'\varepsilon(a1'').$$

for any $a \in H$, then we call S a pre-antipode of H. In addition, if $\sum_a S(a')a''S(a''') = S(a)$ for any $a \in H$, then S is called a Nill's antipode of H. A bimonoidal almost bialgebra H with Nill's antipode is called a Nill's weak Hopf algebra (see [24, 2, 3]). Moreover, the category of modules over an almost bialgebra is investigated in order to become a tensor category under certain conditions. The initial motivation to study Nill's weak Hopf algebras was their connection with the theory of algebra extensions [25, 8]. Nill's weak Hopf algebras also appear naturally in the theory of dynamical deformation of quantum groups and provide a natural framework for the study of dynamical twist in the Hopf algebra [5].

With the motivation of studying the relation between the theory of semigroups and the theory of Hopf algebra, Li gave another kind of generalization of Hopf algebra in [9-10], named also weak Hopf algebra. Let $H = (H, m, u, \Delta, \varepsilon)$ be a bialgebra (see [26]). Then we have the convolution product algebra $\text{Hom}_k(H, H)$ with the multiplication $f * g = m(f \otimes g)\Delta$ for all $f, g \in \text{Hom}_k(H, H)$. By definition, a bialgebra H is called a weak Hopf algebra [9-10] if there exists $T \in \text{Hom}_k(H, H)$ (a weak antipode) such that $id * T * id = id$ and $T * id * T = T$.

A typical example of weak Hopf algebras is the semigroup algebra kS of regular semigroup S over a field k as a generalization of group algebra kG.

It is well known that from every finite dimensional Hopf algebra, a Drinfeld's quantum double can be constructed, which possesses an R-matrix as a solution of the QYBE. However, this solution is invertible. The other motivation for Li to introduce Li's weak Hopf algebras is to construct singular solutions of QYBE. It is interesting that in [9, 11] the author obtained a class of von Neumann's regular solution of the QYBE from a class of weak Hopf algebras.

Li's weak Hopf algebra is always a bimonoidal almost bialgebra, but the weak antipode of a weak Hopf is not usually a pre-antipode. Hence the class of Nill's weak Hopf algebras and the class of Li's weak Hopf algebras cannot be included with each other. In summary, we show their relation using the following diagram:

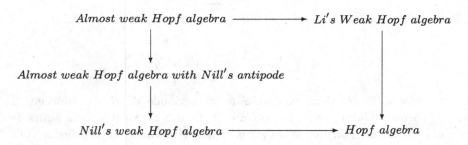

In the present paper, denote the set of all natural numbers by $\mathbf{N}$. We always assume that any algebra is over a field k. The notations and knowledge of Hopf algebra are based on [26]. In the sequel, we mainly restrict our attention to the Li's weak Hopf algebra. So, weak Hopf algebras will always be under the Li's meaning.

First, we recall some elementary algebraic properties which show the relation between weak Hopf algebras and the regular properties of the monoid of its set of group-like elements .

Let S be a semigroup, the set of all idempotents of S is written as $E(S)$. If $S = E(S)$, we call this semigroup S a band. If a band is commutative, it is called a semilattice.

Proposition 0.1. [12] *Let H be a weak Hopf algebra. Denote by $V(id)$ the set of all weak antipodes of H. Define $S \diamond T = S * id * T$ for any $S, T \in V(id)$, then $(V(id), \diamond)$ is a band, where $*$ is just the convolution* *product in $Hom_k(H, H)$.*

Let H be a weak Hopf algebra with antipode T, if T is an anti-algebra morphism and A is a commutative algebra, then $Alg(H, A)$ is a regular monoid with the convolution operation $*$. Dually, if T is an anti-coalgebra morphism and A is a cocommutative coalgebra, then $Coalg(H, A)$ is also a regular monoid with the convolution operation $*$.

Proposition 0.2. [12] *Let $R_k(S) = \{f \in Map(S, k) : dim(kSf) < \infty\}$, then $(R_k(S), \mu, \Delta, \eta, \varepsilon)$ is a bialgebra structure. If S is an invertible monoid, then $R_k(S)$ is a weak Hopf algebra with antipode T satisfying $T(f)(x) = f(x^{-1})$ for any $f \in R_k(S)$, $x \in S$.*

Theorem 0.1. [12] *Let H be a bialgebra, if $E(G(H))$ is a band (resp., semilattice), then the subbialgebra $B = \sum_{g \in E(G(H))} H_g$ is a supplementary band (resp. semilattice) sum of subbiagebras H_g for $g \in E(G(H))$.*

Moreover, if $E(G(H))$ is a semilattice, then $B = \sum_{g \in E(G(H))} H_g$ is regular (resp. strongly regular, weakly regular) if and only if so is H_g for each $g \in E(G(H))$.

When the monoid of group-like elements is a Clifford monoid, a left quasi-module-bialgebra can be obtained from a sum of some irreducible components. Explicitly,

Theorem 0.2. [12] *(i) Assume that H is a bialgebra whose $G = G(H)$ is a Clifford monoid. Then $B = \sum_{g \in E(G(H))} H_g$ is a left kG-module with the structure map $\varphi : kG \otimes B \longrightarrow B$ satisfying $\varphi(g \otimes h) = ghg^{-1}$ for $h \in B, g \in G$; (ii) If moreover, $E(G) \subseteq C(B)$, then B is a left kG-quasi-module-bialgebra.*

Furthermore, the author proved that: if H is a weak Hopf algebra with weak antipode T which is an anti-bialgebra morphism and $G = G(H)$ is a Clifford monoid satisfying $E(G) \subseteq C(B)$ for $B = \sum_{g \in E(G(H))} H_g$. Then B is a left kG-quasi-module-weak Hopf algebra with the structure map $\varphi : kG \otimes B \longrightarrow B$ satisfying $\varphi(g \otimes h) = ghg^{-1}$ for $h \in B, g \in G$.

In the following, we give some results which turn out the role of a weak antipode in the structure of modules or comodules over a weak Hopf algebra.

Proposition 0.3. [12] *Assume H is a weak Hopf algebra with weak antipode T. Define $\lhd$ and $\rhd$ satisfying that $< x \lhd f, y > = < f, T(x)y >$ and $< f \rhd x, y > = < f, yT(x) >$ for $x, y \in H, f \in H^*$. Then*

(i) If T is an anti-algebra morphism, then H^* becomes a left H-module and a right H-module respectively on $\lhd$ and $\rhd$;

(ii) If T is an anti-bialgebra morphism, then H^* becomes a left H^{cop}-module algebra and a right H^{cop}-module algebra respectively on $\triangleleft$ and $\triangleright$.

Proposition 0.4. [12] Assume H is a weak Hopf algebra whose weak antipode T is an anti-bialgebra morphism. Set $\psi_H = \psi : H \longrightarrow H$ with $\psi(x) = \sum_{(x)} xT(x''') \otimes x''$ for any $x \in H$. Then

(i) H is a left H-comodule with its structure map ψ;

(ii) If H is commutative, then H is a left H-comodule algebra;

(iii) If H is cocommutative and T is a right antipode, then H is a left H-comodule coalgebra;

(iv) If H is commutative and cocommutative and T is a right antipode, then H is a left H-comodule right Hopf algebra.

Dually, we can define $\psi'(x) = \sum_{(x)} x'' \otimes T(x')X'''$ so as to construct the structure of a right H-comodule and etc.

Let H and N be both weak Hopf algebras over k with the weak antipodes T_H and T_N, respectively, φ a bialgebra morphism from H to N such that $\varphi T_H = T_N \varphi$. Then φ is called a weak Hopf algebra morphism. Let $\iota : H \otimes H \longrightarrow H \otimes H$ be the twist mapping defined by $\iota(h_1 \otimes h_2) = h_2 \otimes h_1$, for every $h_1, h_2 \in H$ and let $m^{op} : H \otimes H \longrightarrow H$, $\Delta^{cop} : H \longrightarrow H \otimes H$ be the maps defined by $m^{op} = m\iota$ and $\Delta^{cop} = \iota\Delta$ (the opposite multiplication and the opposite comultiplication). If $H = (H, m, u, \Delta, \varepsilon)$ is bialgebras, then $H^{opcop} = (H, m^{op}, u, \Delta^{cop}, \varepsilon)$, $H^{op} = (H, m^{op}, u, \Delta, \varepsilon)$, $H^{cop} = (H, m, u, \Delta^{cop}, \varepsilon)$ are bialgebras [7]. Hence, we have

Proposition 0.5. [9] Let $H = (H, m, u, \Delta, \varepsilon)$ denote a weak Hopf algebra with weak antipode T. Then

(i) $H^{opcop} = (H, m^{op}, u, \Delta^{cop}, \varepsilon, T)$ is a weak Hopf algebra. If moreover, for every $h_1, h_2 \in H$, $T(h_1 h_2) = T(h_2)T(h_1)$ and $\Delta T = \iota(T \otimes T)\Delta$, then $T : H \longrightarrow H^{op,cop}$ is a weak Hopf algebra morphism;

(ii) If T is invertible and $T(h_1 h_2) = T(h_2)T(h_1)$ for any $h_1, h_2 \in H$, then $H^{op} = (H, m^{op}, u, \Delta, \varepsilon, T^{-1})$ and $H^{cop} = (H, m, u, \Delta^{cop}, \varepsilon, T^{-1})$ are weak Hopf algebras with weak antipode T^{-1}. If moreover, $\Delta T = \iota(T \otimes T)\Delta$, then $H^{op} \simeq H^{cop}$ via a weak Hopf algebra isomorphism T.

A weak Hopf algebra H is called a perfect weak Hopf algebra [12] if its weak antipode T is an antibialgebra morphism satisfying $(id * T)(H) \subseteq C(H)$ (the center of H); a coperfect weak Hopf algebra [11] if its weak antipode is an antibialgebra morphism satisfying $\sum h_1 T(h_2) \otimes h_3 = \sum h_2 T(h_3) \otimes h_1$ for any $h \in H$; a biperfect weak Hopf algebra if it is perfect and also coperfect. A finite dimensional weak Hopf algebra H is perfect (resp., coperfect) if and only if its duality H^* is coperfect (resp., perfect). If H is a weak Hopf algebra with a bijective antipode, then H is perfect (resp., coperfect) if and only if H^{op} (resp., H^{cop}) is also perfect (resp., coperfect).

1. Generalized quantum double and singular solutions of QYBE

As is well known, braided Hopf algebras provid solutions of Yang-Baxter equation. The key is how to find enough such interesting Hopf algebras. Drinfeld devised an ingenious method, quantum double construction, which builds a braided Hopf algebra out of any finite dimensional Hopf algebra with invertible antipode. In these classical methods, the given solutions of QYBE are always invertible. In the method through weak Hopf algebras, the original key is to obtain the non-invertible solutions of QYBE. With this motivation, the quantum quasi-double as a generalization of quantum doubles from some certain weak Hopf algebra is introduced so as to obtain such solutions of QYBE. As is well known, a bicrossed product is a fundamental tool for constructing the quantum double of a Hopf algebra; a quasi-bicrossed product plays a similar role for the quantum quasi-double of a certain weak Hopf algebra (in particular, the semigroup algebra of finite Clifford monoid).

Definition 1.1. [9] A pair (X, A) of bialgebras is called quasi-matched if there exist linear maps $\alpha : A \otimes X \longrightarrow X$ and $\beta : A \otimes X \longrightarrow A$ turning X into a left A-quasi-module-coalgebra and turning A into a right X-quasi-module-coalgebra such that if one sets $\alpha(a \otimes x) = a \cdot x$, $\beta(a \otimes x) = a^x$, the following conditions are satisfied:

$$a \cdot (xy) = \sum (a' \cdot x')(a''^{x''} \cdot y),$$

$$a \cdot 1 = \varepsilon(a)1,$$

$$(ab)^x = \sum a^{b' \cdot x'} b'' x'',$$

$$1^x = \varepsilon(x)1,$$

$$\sum a'^{x'} \otimes a'' \cdot x'' = \sum a''^{x''} \otimes a' \cdot x'.$$

for all $a, b \in A$ and $x, y \in X$.

Let (X, A) be a quasi-matched pair of bialgebras. From [9], we know that there exists an almost bialgebra structure on the vector space $X \otimes A$ with unit equal to $1 \otimes 1$ such that its product is given by $(x \otimes a)(y \otimes b) = \sum x(a' \cdot y') \otimes a''^{y''} b$, its coproduct given by $\Delta(x \otimes a) = \sum (x' \otimes a') \otimes (x'' \otimes a'')$ and its counit by $\varepsilon(x \otimes a) = \varepsilon_A(a)\varepsilon_X(x)$ for all $a, b \in A$ and $x, y \in X$. Equipped with this almost bialgebra structure, $X \otimes A$ is called the quasi-bicrossed product of X and A and denoted $X \infty A$. Furthermore, the injective maps $i_X(x) = x \otimes 1$ and $i_A(a) = 1 \otimes a$ from X and from A into $X \infty A$ are bialgebra morphisms. Also $x \otimes a = (x \otimes 1)(1 \otimes a)$ for $a \in A$ and $x \in X$.

Note that for two arbitrary bialgebras X and A with the quasi-bicrossed product $X \infty A$, in general, $X \infty A$ is noncommutative as an algebra; and if either X or A is noncocommutative, $X \infty A$ is noncocommutative as a coalgebra. Hence we can construct many noncommutative and noncocommutive almost bialgebras in the form of the quasi-bicrossed product $X \infty A$.

In particular, consider the case that $A = H$ is a finite dimensional weak Hopf algebra with invertible weak antipode T, then $X = (H^{op})^*$ is also so. Let $H = (H, m, u, \Delta, \varepsilon, T)$ be a finite dimensional cocommutative perfect weak Hopf algebra with T invertible. Then for $X = (H^{op})^* = (H^*, (m^{op})^*, u^*, \Delta^*, \varepsilon^*, (T^{-1})^*)$, the pair (X, H) of bialgebras is quasi-matched with the linear maps α and β given by $\alpha(a \otimes f) = a \cdot f = f(T^{-1}(a'')?a')$ and $\beta(a \otimes f) = a^f = \sum f(T^{-1}(a''')a')a''$ for $a \in H$ and $f \in X$ where $f(T^{-1}(a'')?a')$ means the map $x \longrightarrow f(T^{-1}(a'')xa')$ for any $x \in H$.

Definition 1.2. [9] Let $H = (H, m, u, \Delta, \varepsilon, T)$ be a finite dimensional cocommutative perfect weak Hopf algebra with T invertible. The quasi-bicrossed product $(H^{op})^* \infty H$ of $(H^{op})^*$ and H is called the quantum quasi-double of H, denoted as $D(H)$, i.e., $D(H) = (H^{op})^* \infty H$.

Theorem 1.1. [9] *Let $H = (H, m, u, \Delta, \varepsilon, T)$ be a finite dimensional cocommutative perfect weak Hopf algebra with T invertible. Then the quantum quasi-double $D(H)$ of H is quasi-braided equipped with the quasi-R-matrix $R = \sum_{i \in I} (1 \infty e_i) \otimes (e_i \infty 1) \in D(H) \otimes D(H)$, where $\{e_i\}_{i \in I}$ is a basis of the k-vector space H together with its dual basis $\{e^i\}_{i \in I}$ in $(H^{op})^*$. Hence R is a solution of the QYBE.*

Theorem 1.1 means that one can find some solutions of the YBE built from some almost bialgebras. The importance of a quantum double is its R-matrix as a nonsingular solution of the QYBE. For a quantum quasi-double, in [11] we showed that every quasi-R-matrix R of a quasi-braided almost bialgebra H is a regular (under the von Neumann's meaning) solution of the QYBE: $R_{12}R_{13}R_{23} = R_{23}R_{13}R_{12}$, whose inverse is $\hat{R} = (T \otimes id)(R)R(T \otimes id)(R)$, i.e. $\hat{R}R\hat{R} = \hat{R}$ and $R\hat{R}R = R$. As a result, a regular solution of the classical YBE can be built.

Now we give an example of weak Hopf algebra satisfying Theorem 1.1 by using of a finite Clifford monoid $S = [Y; G_\alpha, \varphi_{\alpha,\beta}]$ (i.e. a strong semilattice of groups G_α). The elementary conclusions on a Clifford semigroup contain that its set of all idempotents $E(S) \subseteq C(S)$ (the center of S), and S is an inverse semigroup. Thus for every $a \in S$ there exists a unique inverse a^{-1} such that $(a^{-1})^{-1} = a$ and $(ab)^{-1} = b^{-1}a^{-1}$. Suppose $S = \{s_1, s_2, \cdots, s_n\}$ is finite, then kS becomes a finite dimensional cocommutative weak Hopf algebra with weak antipode $T : T(a) = a^{-1}$ for $a \in S$. We have $T = T^{-1}$. It is easy to see that kS satisfies the condition of the result above. Thus, we get the quantum quasi-double $D(kS)$ of kS, briefly, denote $D(S) = D(kS)$ and call $D(S)$ the quantum quasi-double of the Clifford monoid S. Thus $D(S)$ is quasi-braided equipped with the quasi-R-matrix $R = \sum_{i=1}^{n}(1 \infty s_i) \otimes (s_i^* \infty 1) \in D(S) \otimes D(S)$. It follows that R is a solution of the QYBE, where $s_i^* : kS \longrightarrow k$ satisfying $s_i^*(s_j) = \delta_{ij}$. As is well known, a Clifford semigroup is a generalization of a group. Hence, this result generalize the method of constructing the solution of the QYBE from a group algebra to a semigroup algebra.

As a natural generalization of quantum double from a group G in [15], we obtain the algebraic structure of $D(S)$ and the relation to some quantum doubles from groups G_α for a clifford group $S = \cup_{\alpha \in L} G_\alpha$.

Theorem 1.2. [15] *(Structure Theorem) For a finite Clifford monoid $S = \cup_{\alpha \in L} G_\alpha$ as a strong semilattice L of groups G_α, the quantum double $D(S)$ is a direct sum of right ideals $Q_S(G_\alpha)$, $\alpha \in L$, where*

(i) every $Q_S(G_\alpha)$ is a supplementary semilattice sum of subrings $D(G_\alpha, G_\beta)$ for $\beta \in L$ and is a coalgebra with comultiplication Δ satisfying $\Delta((\phi_a \infty x)(\phi_b \infty y)) = \Delta(\phi_a \infty x)\Delta(\phi_b \infty y)$ for any x, $y \in S$ and a, $b \in G_\alpha$;

(ii) meantime, $Q_S(G_\alpha) = N_S(G_\alpha) \oplus B_S(G_\alpha)$ where $N_S(G_\alpha) = \sum_{\beta \in L, \beta \not\geq \alpha} D(G_\alpha, G_\beta)$ is a null right ideal of $D(S)$ and is a subcoalgebra and ideal of $Q_S(G_\alpha)$, $B_S(G_\alpha) = \sum_{\beta \in L, \beta \geq \alpha} D(G_\alpha, G_\beta)$ is a sub-Hopf algebra of $Q_S(G_\alpha)$ with $N_S(G_\alpha)B_S(G_\alpha) = 0$ and $B_S(G_\alpha)N_S(G_\alpha) \subseteq N_S(G_\alpha)$;

(iii) $D(G_\alpha, G_\beta)$ are subcoalgebras of $Q_S(G_\alpha)$. If $\alpha \not\leq \beta$, $D(G_\alpha, G_\beta)$ is a null subring. If $\alpha \leq \beta$, $D(G_\alpha, G_\beta)$ is a Hopf algebra. If $\alpha = \beta$, then $D(G_\alpha, G_\alpha) = D(G_\alpha)$, which means that every quantum double $D(G_\alpha)$ is a direct sum component of $D(S)$.

Moreover, in the same paper, it is shown that for a finite Clifford monoid S, its quantum double $D(S)$ over a field k is semisimple (resp. regular) if and only if S is a group and the characteristic p of k does not divide the order $|S|$ of S.

Since the quantum quasi-double of a Clifford monoid S is a generalization of the quantum double of a group, for $S = [Y; G_\alpha, \varphi_{\alpha,\beta}]$, by using of the relation of $D(S)$ and $D(G_\alpha)$, it will be possible to generalize some properties of the quantum double $D(G_\alpha)$ to $D(S)$. If one says that a group expresses the inner symmetry of a physical system, we believe that a Clifford monoid, as a strong semilattice, can express a relationship among some different physical systems through a certain operation such that it is possible to investigate integrally the set of different systems from each only system. We have known from G. Mason [23] that quantum doubles of groups play an important role in conformal field theory, in particular, in the theory of holomorphic orbifolds. We try to find the similar role of quantum quasi-doubles of Clifford monoids.

Analogously to Hopf algebras, weak Hopf pair was displayed in [13], which are generalizations of the Hopf pairs introduced by Takeuchi [27]. Using the weak Hopf skew pairs, one type of quasi-bicrossed products, which lie between general quasi-bicrossed products and quantum quasi-doubles, are constructed when one of the two weak Hopf algebras in the product is cocommutative.

Definition 1.3. [13] (i) Suppose that X and A are weak Hopf algebras with weak antipodes S_X and S_A. We call (X, A) a weak Hopf pair, if there exists a non-singular bilinear form $<,>$ from $X \otimes A$ to k satisfying that

$$< x, ab > = \sum < x', a >< x'', b >, \tag{1}$$

$$< x, 1_A > = \varepsilon(x), \tag{2}$$

$$< xy, a > = \sum < x, a' >< y, a'' >, \tag{3}$$

$$< 1_X, a > = \varepsilon(a), \tag{4}$$

$$< S_X(x), a > = < x, S_A(a) > \tag{5}$$

where $x, y \in X$, $a, b \in A$.

(ii) In (i), moreover, if (1) and (5) are replaced with the following

$$< x, ab > = \sum < x'', a >< x', b >, \tag{6}$$

$$< S_X(x), a >=< x, S_A^{-1}(a) > . \qquad (7)$$

we call (X, A) a weak Hopf skew pair.

For two more general perfect weak Hopf algebras X and A with weak antipodes S_X and S_A respectively, suppose that A is cocommutative, S_A is invertible and (X, A) is a weak Hopf skew pair. Then (X, A) is a quasi-matched pair of bialgebras with $a \triangleright x = \sum < x' S_X(x'''), a > x''$, $a \triangleleft x = \sum < x, S_A^{-1}(a''')a' > a''$ so as to get a quasi-bicrossed product $X \infty A$, denoted as $D(X, A)$.

As a continuous work of [13], the authors in [20] constructed the quasi-bicrossed product of the weak Hopf skew pair corresponding to the case where both weak Hopf algebras in the product are noncocommutative. Let X and A be two perfect weak Hopf algebras with weak antipodes S_X and S_A respectively, assume that S_A is invertible and (X, A) is a weak Hopf skew pair. Then (X, A) is a quasi-matched pair of bialgebras with $a \triangleright x = \sum < x' S_X(x'''), a > x''$, $a \triangleleft x = \sum < x, S_A^{-1}(a''')a' > a''$ so as to get a quasi-bicrossed product $X \infty A$, denoted as $D(X, A)$.

We now consider the particular case that $A = H$ is a finite dimensional weak Hopf algebra. Let $H = (H, m, u, \Delta, \varepsilon, T)$ be a finite dimensional biperfect weak Hopf algebra with invertible T. Then the multiplication in $D(H) = H^{*cop} \infty H$ is given by $(f \infty a)(g \infty b) = \sum fg(T^{-1}(a''')?a') \infty a''b$ for $f, g \in H^{*cop}, a, b \in H$.

Theorem 1.3. [20] *Let $H = (H, m, u, \Delta, \varepsilon, T)$ be a finite dimensional biperfect weak Hopf algebra with invertible T. Then the quantum double $D(H)$ of H is quasi-braided equipped with a quasi-R-matrix $R = \sum_{i \in I}(1 \infty e_i) \otimes (e_i \infty 1) \in D(H) \otimes D(H)$, where $\{e_i\}_{i \in I}$ is a basis of the k-vector space H together with its dual basis $\{e^i\}_{i \in I}$ in $(H^{cop})^*$. Hence R is a solution of the QYBE.*

Comparing this result with Theorem 1.1, it is interesting to note that the cocommutativity is replaced by biperfect condition here.

As in the case of Hopf algebras [21], Wang and Ma constructed a weak bicrossed coproduct $X \infty_R A$ [28] by means of good regular R-matrices of the weak Hopf algebras X and A. Using this, they provided a new framework of obtaining singular solutions of the QYBE by constructing weak quasi-triangular structures over $X \infty_R A$ when both X and A admit a weak quasi-triangular structure.

Let X and A be almost bialgebras. Let X be a left A-quasi-comodule algebra with comodule structure map: $\Delta_X : X \longrightarrow A \otimes X$, $\Delta_X(x) =$

$\sum x_A \otimes x_X$ for all $x \in X$, and let A be a right X-quasi-comodule algebra with comodule structure map: $\delta_A : A \longrightarrow A \otimes X$, $\delta_A(a) = \sum a_A \otimes a_X$ for all $a \in A$. We set $T : X \otimes A \longrightarrow A \otimes X$, $T(x \otimes a) = \Delta_X(x)\delta_A(a) = \sum x_A a_A \otimes x_X a_X$ and $F : A \otimes X \longrightarrow A \otimes X$, $F(a \otimes x) = \delta_A(a)\Delta_X(x) = \sum a_A x_A \otimes a_X x_X$

Define $X^{\Delta} \infty_\delta A = X \otimes A$ as a tensor product algebra, but with a comultiplication and a counit as follows: $\tilde{\Delta} = (X \otimes T \otimes A)(\Delta_X \otimes \Delta_A)$, $\tilde{\varepsilon} = \varepsilon_X \otimes \varepsilon_A$. $X^{\Delta} \infty_\delta A = X \otimes A$ is an almost bialgebra if and only if the following conditions are satisfied: (i) $(A \otimes \Delta_X)\Delta_X = (T \otimes X)(X \otimes \Delta_X)\Delta_X$

(ii) $(A \otimes \varepsilon_X)\Delta_X = 1_A \varepsilon_X$

(iii) $(\Delta_A \otimes X)\delta_A = (A \otimes T)(\delta_A \otimes A)\Delta_A$

(iv) $(\varepsilon_A \otimes X)\delta_A = 1_X \varepsilon_A$

(v) $T = F\iota$.

Furthermore, if X and A are weak Hopf algebras with antipodes S_X and S_A respectively, then $X^{\Delta} \infty_\delta A$ is also a weak Hopf algebra with the antipode given by $S = (S_X \otimes S_A)\iota F\iota$. The above defined almost bialgebra $X^{\Delta} \infty_\delta A$ is called a quasi-bicrossed coproduct.

Theorem 1.4. [29] *Let $X^{\Delta} \infty_\delta A$ be a quasi-bicrossed coproduct. If $X^{\Delta} \infty_\delta A$ admits a weak quasitriangular structure, then X and A admit a weak quasitriangular structure, and there is a good regular R-matrix $R \in X \otimes A$ of (X, A) with inverse $\overline{R}$ such that $\Delta_X(x) = \tau(R)(A \otimes x)\tau(\overline{R})$, and $\delta_A(a) = \tau(R)(a \otimes X)\tau(\overline{R})$, i.e., $X^{\Delta} \infty_\delta A = X \infty A$.*

As we know, weak Hopf algebra plays an important role in the construction of noninvertible solutions of YBE. On the other hand, there is a tight relation between weak Hopf algebra and regular monoid. Hence, it is necessary to find more nontrivial weak Hopf algebras in order to study these two aspects deeply. For this aim, Li and Cao [16] built a so-called semilattice graded weak Hopf algebra $H = \oplus_{\alpha \in Y} H_\alpha$ from a family of Hopf algebras $\{H_\alpha\}$.

A weak Hopf algebra H with weak antipode T is called a semilattice graded weak Hopf algebra if $H = \oplus_{\alpha \in Y} H_\alpha$ is a semilattice grading sum where H_α is a weak subHopf algebras of H which is a Hopf algebra with antipodes $T|_{H_\alpha}$ for each $\alpha \in Y$ and there homomorphisms of Hopf algebras $\varphi_{\alpha,\beta}$ from H_α to H_β if $\alpha\beta = \beta$, such that for $a \in H_\alpha$ and $b \in H_\beta$, the multiplication $a * b$ in H can be given by $a * b = \varphi_{\alpha,\alpha\beta}(a)\varphi_{\beta,\alpha\beta}(b)$. An example of semilattice graded weak Hopf algebra is just Clifford moniod algebra.

In [22], the notion of self-dual graded weak Hopf algebra and self-dual semilattice graded weak Hopf algebra are introduced. Moreover, if $H =$

$\oplus H_{n\geq 0}H_n$ is graded and each H_n is finite dimensional, then H is called a locally finite graded weak Hopf algebra. A semilattice graded weak Hopf algebra $H = \oplus_{\alpha \in Y}H_\alpha$ is called locally finite if it is locally finite as weak Hopf algebra. Let $H = \oplus_{n\geq 0}H_n$ be a graded weak Hopf algebra such that H is locally finite, then H is called self-dual if $H \simeq H^{gr} = \oplus_{n\geq 0}H_n^*$ as a graded weak Hopf algebra.

The following gives a characterization for H to be self-dual.

Theorem 1.5. [22] *Let $H = \oplus_{\alpha \in Y}H_\alpha$ be a semilattice graded weak Hopf algebra such that each $H_\alpha = \oplus_{n\geq 0}H_{n,\alpha}$ is locally finite $\mathbf{N}$-graded Hopf subalgebra. Further let $A = \oplus_{\alpha \in Y}H_{0,\alpha} = H_0$ and $M = \oplus_{\alpha \in Y}H_{1,\alpha} = H_1$, then H is self-dual if and only if the following conditions hold:*

(i) $A \cong A^*$ as a semilattice graded weak Hopf algebra;

(ii) $M \cong M^*$ as a semilattice graded weak Hopf bimodule over A, using identification of A and A^* given in (i);

(iii) $H \cong A[M]$.

The related quantum G-double $D'(H)$ for H is commutative is constructed in [16]. For a semilattice graded weak Hopf algebra $H = \oplus_{\alpha \in Y}H_\alpha$, we define a product $D'(H) = H^{op*}\infty'H$ from H by

$$(f\infty'a)(g\infty'b) = fg(1_{H_\alpha}?)\infty'ab$$

for $f \in H_{\alpha_1}^{op*}, g \in H_{\alpha_2}^{op*}, a \in H_{\beta_1}, b \in H_{\beta_2}$, where $fg(1_{H_\alpha}?)$ means the morphism, $x \longmapsto f(x')g(1_{H_\alpha}x'')$. We call $D'(H)$ the quantum G-double of H. Thus, we get $D'(H) = \oplus_{\alpha,\beta \in Y}(H_\alpha^* \otimes H_\beta)$ as linear space. Denote $D'(H_\alpha, H_\beta) = H_\alpha^{op*} \otimes H_\beta$ and $Q_{H_\alpha} = H_\alpha^* \otimes H_\beta = \oplus_{\beta \in Y}D'(H_\alpha, H_\beta)$. Then, as linear space, $D'(H) = \oplus_{\beta \in Y}Q_{H_\alpha} = \oplus_{\alpha,\beta \in Y}D'(H_\alpha, H_\beta)$. Therefore, based on [16], a regular solution of the YBE can be obtained as follows.

Theorem 1.6. [16] *For a commutative semilattice graded weak Hopf algebra $H = \oplus_{\alpha \in Y}H_\alpha$, the quantum G-double $D'(H)$ defined above is a noncommutative and noncocommutative almost bialgebra with regular R-matrix $R = \sum_{\alpha \in Y}\sum_{a_\alpha \in B_\alpha}(\varepsilon_{H_\alpha}\infty'a_\alpha) \otimes (\phi_{a_\alpha}\infty'1_{H_\alpha})$, but it is usually not quasi-triangular although $(\Delta \otimes id_H)(R) = R_{13}R_{23}$ and $(id_H \otimes \Delta)(R) = R_{13}R_{12}$ hold.*

At last in this section, we give a result about the relation between the representation category of a quantum quasi-double from a weak Hopf algebra H in a special case and the related crossed H-bimodule category.

Theorem 1.7. [13] *Suppose H is a finite dimensional cocommutative perfect weak Hopf algebra with invertible weak antipode T. Then,*

(i) *Any left $D(H)$-module has a natural structure as a crossed H-bimodule;*

(ii) *For a croosed H-bimodule V, if*

$$\sum_{(a)(\beta)} T^{-1}(a''')a''\beta_H \otimes a'\beta_V = \sum_{(\beta)} \beta_H \otimes a\beta_V$$

for all $a \in H$ and $\beta \in V$, then V has a natural structure as a left $D(H)$-module.

2. Constructing quantum enveloping algebras via weak Hopf structures

We believe that for the furthermore understanding of weak Hopf algebras as well as having some insight into their applications, it is important to find various examples from some different areas. In this section we mainly list some typical examples of quantum enveloping algebras, which have weak Hopf algebras structures.

Li and Duplij [18] constructed the typical weak Hopf algebras from the quantum algebra $sl_q(2)$ by replacing the set of grouplike elements of $sl_q(2)$ by the set of some regular monoid, that is, exchanging its invertibility $KK^{-1} = 1$ to the regularity $K\overline{K}K = K$. This leads to the weak quantum algebras $U_q^w = w(sl_q(2))$ and $U_q^v = v(sl_q(2))$. In detail, the algebra $w(sl_q(2))$ is generated by four variables (Chevalley generators) $E_w, F_w, K_w, \overline{K_w}$ with the following relations:

$$K_w\overline{K_w} = \overline{K_w}K_w = J_w, \quad K_w\overline{K_w}K_w = K_w$$

$$\overline{K_w}K_w\overline{K_w} = \overline{K_w}, \quad K_wE_w = q^2E_wK_w$$

$$\overline{K_w}E_w = q^{-2}E_w\overline{K_w}, \quad K_wF_w = q^{-2}F_wK_w$$

$$\overline{K_w}F_w = q^2F_w\overline{K_w}, \quad E_wF_w - F_wE_w = \frac{K_w - \overline{K_w}}{q - q^{-1}}.$$

We call such a quantum enveloping algebra as a weak quantum algebra.

The algebra $v(sl_q(2))$ is generated by four variables (Chevalley generators) $E_v, F_v, K_v, \overline{K_v}$ with the relations

$$K_v\overline{K_v} = \overline{K_v}K_v = J_v,$$

$$K_v\overline{K_v}K_v = K_v, \quad \overline{K_v}K_v\overline{K_v} = \overline{K_v},$$

$$K_vE_v\overline{K_v} = q^2E_v, \quad K_vF_v\overline{K_v} = q^{-2}F_v,$$

$$E_v J_v F_v - F_v J_v E_v = \frac{K_v - \overline{K_v}}{q - q^{-1}}.$$

We call $v(sl_q(2))$ the J-weak quantum algebra.

In [17], it was proved that $w(sl_q(2))$ possesses a quasi-R-matrix which is a singular solution of the QYBE, with a parameter q.

One can find the connection between $U_q^w = w(sl_q(2))$, $U_q^v = v(sl_q(2))$ and the quantum algebra $sl_q(2)$. Firstly, $w(sl_q(2))/(J_w - 1) \simeq sl_q(2)$, $v(sl_q(2))/(J_v - 1) \simeq sl_q(2)$.

As a continuous work of [17], Cheng and Li [4] proved that the algebraic structure of $w(sl_q(2))$ can be written as the direct sum of $U_q(sl_2)$ and an algebra of polynomials, that is, $w(sl_q(2)) = w(sl_q(2))J_w \oplus w(sl_q(2))(1 - J_w)$, where $w(sl_q(2))J_w$ is isomorphic to $U_q(sl_2)$ as Hopf algebra, and $w(sl_q(2))(1 - J_w)$ is isomorphic to polynomials algebra $k[x, y]$.

Furthermore, the coalgebraic structure of the weak Hopf algebras $w(sl_q(2))$ and $v(sl_q(2))$ corresponding to $U_q(sl_2)$ are classified by their Ext quivers in [4].

In [1], Aizawa and Isaac proposed some minor adjustments to the weak quantum algebras given as above and extended their construction to the cases of the other known Hopf algebras $U_q(sl_n)$ and the Sweedler's Hopf algebra. As a consequence, they obtained the general weak quantum algebras $\omega sl_q(n)$ for any n. The relations of $\omega sl_q(n)$ satisfied by the generators are as follows, for all i, j unless specified otherwise:

$$J = K_i \overline{K_i} = \overline{K_i} K_i, \quad K_i J = J K_i = K_i, \quad \overline{K_i} J = J \overline{K_i} = \overline{K_i}$$

$$K_i K_j = K_j K_i, \quad K_i \overline{K_j} = \overline{K_j} K_i, \quad \overline{K_i}\, \overline{K_j} = \overline{K_j}\, \overline{K_i},$$

$$E_i F_j - F_j E_i = \delta_{ij} \frac{K_i - \overline{K_i}}{q_i - q_i^{-1}},$$

$$E_i^2 E_{i \pm 1} - (q + q^{-1}) E_i E_{i \pm 1} E_i + E_{i \pm 1} E_i^2 = 0$$

$$F_i^2 F_{i \pm 1} - (q + q^{-1}) F_i F_{i \pm 1} F_i + F_{i \pm 1} F_i^2 = 0$$

$$E_j E_i = E_i E_j, \quad F_j F_i = F_i F_j, \quad |i - j| \geq 2.$$

We also need to specify the relations between the E_i and the K_j, for example. Let a_{ij} denote the Cartan matrix for $sl(n)$, $a_{ii} = 2$, $a_{i,i \pm 1} = 2$ and zero otherwise. If E_i satisfies

$$K_j E_i = q_i^{a_{ij}} E_i K_j, \quad E_i \overline{K_j} = q_i^{a_{ij}} \overline{K_j} E_i, \quad \forall j,$$

we say E_i to be of the type 1. Moreover, if E_i satisfies

$$K_j E_i \overline{K_j} = q_i^{a_{ij}} E_i, \quad \forall j,$$

we say E_i to be of the type 0. The same convention holds for F_i by replacing E_i with F_i and a_{ij} with $-a_{ij}$ in the above relations. Notice also that J is defined for all i, so, for example, $J = K_i \overline{K_i} = \overline{K_j} K_j$, $i \neq j$. Furthermore, the weak Hopf algebra structure on $\omega sl_q(n)$ can be decomposed into a direct sum of the original Hopf algebra $U_q(sl_n)$ with some other subalgebra.

Let $U_q(g)$ be the quantized enveloping algebra associated to the finite dimensional semisimple Lie algebra g defined by [Lu], that is, the algebra $U_q(g)$ is generated by 4n generators $E_i, F_i, K_i^{\pm 1}$ ($i = 1, 2, \ldots, n$) and satisfying the following relations:

$$K_i K_i^{-1} = K_i^{-1} K_i = 1, \quad K_i K_j = K_j K_i,$$

$$K_i E_j = q_i^{a_{ij}} E_j K_i, \quad K_i F_j = q_i^{-a_{ij}} F_j K_i,$$

$$E_i F_j - F_j E_i = \delta_{ij} \frac{K_i - K_i^{-1}}{q_i - q_i^{-1}},$$

$$\sum_{k=0}^{1-a_{ij}} (-1)^k \begin{bmatrix} 1 - a_{ij} \\ k \end{bmatrix}_{q_i} E_i^{1-a_{ij}-k} E_j E_i^k = 0, \quad i \neq j,$$

$$\sum_{k=0}^{1-a_{ij}} (-1)^k \begin{bmatrix} 1 - a_{ij} \\ k \end{bmatrix}_{q_i} F_i^{1-a_{ij}-k} F_j F_i^k = 0, \quad i \neq j.$$

Yang [30] replaced $\{K_i, K_i^{-1}\}$ by $\{K_i, \overline{K_i}\}$ for any $1 \leq i \leq n$ and introduced a projector J such that

$$J = K_i \overline{K_i} = \overline{K_i} K_i,$$

$$J K_i = K_i J = K_i, \quad J \overline{K_i} = \overline{K_i} J = \overline{K_i}.$$

To generalize other relations of the definition, the authors introduced some notations as follows: if E_i satisfies

$$K_j E_i = q_i^{a_{ij}} E_i K_j, \quad E_i \overline{K_j} = q_i^{a_{ij}} \overline{K_j} E_i, \quad \forall j,$$

which is said to be of type 1. And, if E_i satisfies

$$K_j E_i \overline{K_j} = q_i^{a_{ij}} E_i, \quad \forall j,$$

which is said to be of type 0. The same convention holds for F_i by replacing E_i with F_i and a_{ij} with $-a_{ij}$ in the above relations. We write d in terms of its binary expansion,

$$d = (k_1, \cdots, k_n | \overline{k_1}, \cdots, \overline{k_n})$$

where the bar separates the values representing the $\{E_i\}_{1 \le i \le n}$ and $\{F_i\}_{1 \le i \le n}$, and where the $\{k_i\}_{1 \le i \le n}$ and $\{\overline{k_i}\}_{1 \le i \le n}$ have values of either 0 or 1. Accordingly, we say $(E_1, \cdots, E_n | F_1, \cdots, F_n)$ to be of type d in an obvious sense. Explicitly,

Definition 2.1. [30] The algebra $m_q^d(g)$ is generated by the E_i, F_i, K_i, $\overline{K_i}$ and J with the following relations for all $1 \le i, j \le n$,

$$J = K_i \overline{K_i} = \overline{K_i} K_i,$$

$$K_i J = J K_i = K_i, \quad \overline{K_i} J = J \overline{K_i} = \overline{K_i},$$

$$K_i K_j = K_j K_i, \quad K_i \overline{K_j} = \overline{K_j} K_i, \quad \overline{K_i}\, \overline{K_j} = \overline{K_j}\, \overline{K_i},$$

$$(E_1, \cdots, E_n | F_1, \cdots, F_n) \text{ is of type } d,$$

$$E_i F_j - F_j E_i = \delta_{ij} \frac{K_i - \overline{K_i}}{q_i - q_i^{-1}},$$

$$\sum_{k=0}^{1-a_{ij}} (-1)^k \begin{bmatrix} 1 - a_{ij} \\ k \end{bmatrix}_{q_i} E_i^{1-a_{ij}-k} E_j E_i^k = 0, \quad if \ \ i \ne j,$$

$$\sum_{k=0}^{1-a_{ij}} (-1)^k \begin{bmatrix} 1 - a_{ij} \\ k \end{bmatrix}_{q_i} F_i^{1-a_{ij}-k} F_j F_i^k = 0, \quad if \ \ i \ne j,$$

The algebra $m_q^d(g)$ is said to be a d-type weak quantum algebra associated to the Lie algebra g, where $(a_{ij})_{n \times n}$ is the Cartan matrix corresponding to g.

There are 2^{2n} possible weak quantum algebras $m_q^d(g)$ corresponding to the sequence d in total.

The algebra $m_q^d(g)$ is a noncommutative and noncocommutative weak Hopf algebra. One can find it has closely connection with the quantized enveloping algebra $U_q(g)$. Explicitly, we have $m_q^d(g)/ < J - 1 > \simeq U_q(g)$, and as algebras $m_q^d(g) = m_q^d(g)J \oplus m_q^d(g)(1 - J)$ with $m_q^d(g)J \simeq U_q(g)$ as Hopf algebras.

A generalized Kac-Moody algebra g can be regarded as a Kac-Moody algebra with imaginary simple roots, which is determined by a Borcherds-Cartan matrix $A = (a_{ij})_{(i,j) \in (I \times I)}$, where I is a set. Using the techniques of [AI] and [Y], the authors in [32] naturally constructed the more general weak Hopf algebra as a new weak quantum algebra from the quantum algebra $U_q(g)$ associated to the generalized Kac-Moody algebra g. By weakening the invertibility to regularity, Wu in [33] replaced $\{K_i, K_i^{-1}, D_i, D_i^{-1}\}$ by $\{K_i, \overline{K_i}, D_i, \overline{D_i}\}$ for any $i \in I$ and introduced a projector J such that

$$J^m = J^{m-1} = K_i \overline{K_i} = \overline{K_i} K_i = D_i \overline{D_i} = \overline{D_i} D_i, \ (m \in \mathbf{N})$$

$$JK_i = K_i J = K_i, \ J\overline{K_i} = \overline{K_i} J = \overline{K_i}, \ JD_i = D_i J = D_i, \ J\overline{D_i} = \overline{D_i} J = \overline{D_i}.$$

and the other generators satisfying the similar relations as above. Consequently, more nontrivial examples of weak Hopf algebras were given, denoted as $wU_q^d(g)$.

It is interesting to consider how such weak quantum algebras $wU_q^d(g)$ will change if m is given as some various positive integers.

In [31] and [34], the authors extended the above techniques to the super-object based on quantum superalgebra $U_q(g)$, where g is Borcherds super-algebra. This means one can give some weak Hopf superalgebras as new weak quantum algebras which supplies a new approach to Borcherds su-peralgebras.

Note that in the above discussions, it is always defined that $J = K_i \overline{K_i} = \overline{K_i} K_i$ for any various $i \in \mathbf{N} \setminus \{0\}$. Our question is if we let $J_i = K_i \overline{K_i} = \overline{K_i} K_i$ such that J_i are different for the various i, what will happen for the structures of the weak quantum algebras given above and the related applications?

In [16], the singular solution of QYBE can be obtained from the weak quantum quantum algebra $w(sl_q(2))$. So, the other question we want to consider is how to obtain some corresponding singular solutions of the QYBE from weak quantum algebras given in [1,28, 30-32].

3. Generalizations of tensor category from weak Hopf algebras

As is well known, the representation category of a Hopf algebra is a tensor category (with duality) which plays an important role in the theory of quantum groups, knots and etc. The importance is natural to be understood of the analogues of tensor category from representations of some various classes of weak Hopf algebras.

A pre-tensor category $\mathcal{C} = (\mathcal{C}, \otimes, a)$ is a category which is equipped with a tensor product $\otimes : \mathcal{C} \times \mathcal{C} \longrightarrow \mathcal{C}$ with an associativity constraint a such that the Pentagon Axiom is satisfied (see [14]). The pre-tensor category is said to be strict if the associativity a is an identity of the category $\mathcal{C}$. A quasi-braiding c in $\mathcal{C}$ is a commutativity quasi-constraint satisfying the Hexagon Axiom (see [14]). A quasi-braided pre-tensor category $(\mathcal{C}, \otimes, a, c)$ is a pre-tensor category with a quasi-braiding c. A general Yang-Baxter operator (see [14]) is given as the categorical version of a (singular) solutions of QYBE.

The following is an example of quasi-braided pre-tensor category from a weak Hopf algebra.

Example 3.1. [14] Let $H = (H, \mu, 1, \Delta, \varepsilon, T)$ be a finite dimensional cocommutative perfect weak Hopf algebra with an invertible weak antipode T. By [9], the quantum quasi-double $D(H) = (H^{op})^* \infty H$ of H (as a generalization of the Drinfeld's quantum double of a Hopf algebra) is a quasi-braided almost bialgebra equipped with the universal quasi-R-matrix $R = \sum_{i \in I} (1 \infty e_i) \otimes (e^i \infty 1)$, where $\{e_i\}_{i \in I}$ is a basis of the K-vector space H together with its dual basis $\{e^i\}_{i \in I}$ in $(H^{op})^*$. Hence $D(H)$-Mod is a quasi-braided pre-tensor category with a quasi-braiding c by $c_{V,W}(v \otimes w) = \tau_{V,W}(R(v \otimes w))$ where V and W are $D(H)$-modules and $v \in V, w \in W$.

Define a category $\widehat{\mathcal{B}}^+$ whose objects are finite sequence of + signs without the empty sequence 0, and whose morphisms consist of the elements of all right braid monoids. Obviously, $\widehat{\mathcal{B}}^+$ is a strict discrete pre-tensor subcategory of $\mathcal{B}^+$. We call $\widehat{\mathcal{B}}^+$ the strict right braid category.

For two pre-tensor categories $\mathcal{C}$ and $\mathcal{D}$, denote by $PTens(\mathcal{C}, \mathcal{D})$ the category whose objects are all pre-tensor functors (see [14]) from $\mathcal{C}$ to $\mathcal{D}$. Denote $GYB(\mathcal{C})$ the category of all general Yang-Baxter operators on a pre-tensor category $\mathcal{C}$. The main results in [14] are that (i) $PTens(\widehat{\mathcal{B}}^+, \mathcal{C}) \cong GYB(\mathcal{C})$ (an equivalence of categories), where $\widehat{\mathcal{B}}^+$ is the strict right braid category; (ii) For any quasi-braided pre-tensor category $\mathcal{C}$, $QBr(\widehat{\mathcal{B}}^+, \mathcal{C}) \cong \mathcal{C}$ (an equivalence of categories). They mean indeed two various universality properties of the strict right braid category $\widehat{\mathcal{B}}^+$.

As a matter of fact, the structure of the center of a braided tensor category embodies a categorical mode of a quantum double. Hence at the end of [14], the so-called general center $GZ(\mathcal{C})$ of a quasi-braided pre-tensor category $\mathcal{C}$ was built so as to give a categorical mode of a quantum quasi-double. Explicitly, we have

Theorem 3.1. [14] *Let H be a finite dimensional cocommutative perfect weak Hopf algebra with invertible antipode T. Assume*

$$\sum_{(v)(a)} T^{-1}(a''')a''v_H \otimes a'v_V = \sum_{(v)} v_H \otimes av_V$$

for any left H-Mod and right H-comodule V and all $a \in H$, $v \in V$. Then the general center $GZ(H\text{-Mod})$ (with $c_{X,I} = r_X \cdot l_X^{-1}$ for any object X of H-Mod) and the quasi-braided pre-tensor category $D(H)$-Mod are pre-tensor equivalent through a quasi-braided strict pre-tensor functor F.

Recently in [18], the authors generalized tensor category $\mathcal{C} = (\mathcal{C}, \otimes, I, a, l, r)$ by weakening the natural isomorphisms l, r, that is, exchanging the natural isomorphism $ll^{-1} = rr^{-1} = id$ into regular natural transformations $l\bar{l}l = l$, $r\bar{r}r = r$ with some other conditions so as to define the notion of weak tensor category as the categorical version of representations of weak bialgebras and Nill's weak Hopf algebras. The unit I in weak tensor category plays a similar role just as in tensor category. Every weak tensor category is weak tensor equivalent to a strict one. Moreover, the concept of a regular braided weak tensor category is introduced with its regular but not invertible braiding. Finally, through the following diagrams we describe the relations among some generalizations of bialgebras and Hopf algebras, the related quantum doubles and their representation category as the analogues of tensor categories. Denote pre-tensor categories by PTC, weak tensor categories by WTC, tensor categories by TC, almost bialgebras by AB, weak bialgerbras by WB, bialgebras by B, weak Hopf algebras by WHA, Nill's weak Hopf algebras by $NWHA$, Hopf algebras by HA, and quantum double by QD. Then we have

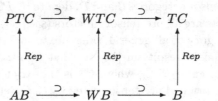

$$
\begin{array}{ccccc}
PTC & \longrightarrow & WTC & \longrightarrow & TC \\
\uparrow Rep & & \uparrow Rep & & \uparrow Rep \\
AB & \longrightarrow & WB & \longrightarrow & B
\end{array}
$$

$$
\begin{array}{ccccc}
quasi-braided\ PTC & \longrightarrow & regular\ braided\ WTC & \longrightarrow & braided\ TC \\
\uparrow Rep & & \uparrow Rep & & \uparrow Rep \\
QD\ of\ WHA & \longrightarrow & QD\ of\ NWHA & \longrightarrow & QD\ of\ HA
\end{array}
$$

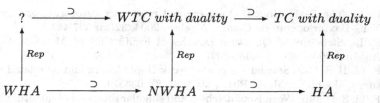

where ? means that so far we don't know how to define the correspondent concept of the duality in PTC so as to make the diagram to be commutative completely.

Acknowledgments

Project supported by the National Natural Science Foundation of China (No. 10871170), the Natural Science Foundation of Zhejiang Province of China (No. D7080064), and the Natural Science Foundation of Heilongjiang Province of China (No. A200906).

References

1. N. Aizawa, P. S. Isaac, Weak Hopf algebras corresponding to $U_q(sl_n)$, J. Math. Phys., 2003, 44(11).
2. G. Böhm, F. Nill, K. Szlachányi, Weak Hopf algebras I. Integral theory and C^*-structure, J. Algebra, 1999, 221: 385-438.
3. G. Böhm, K. Szlachányi, Weak Hopf algebras. II. Representation theory, dimensions, and the Markov trace, J. Algebra, 2000, 233(1): 156-212.
4. D. M. Cheng, F. Li, The structure of weak Hopf algebras corresponding to $U_q(sl_2)$, Comm. Algebra, 2009, 37: 729-742.
5. P. Etingof, D. Nikshych, Dynamical quantum groups at roots of 1, Duke Math., 2001, 108(1): 135-168.
6. T. Hayashi, An algebra related to the fusion rules of Wess-Zumino Witten Models, Lett. Math. Phys., 1991, 22: 291-296.
7. C. Kassel, Quantum Groups, Springer-Verlag, New York, 1995.
8. L. Kadison, D. Nikshych, Frobenius extensions and weak Hopf algebra, J. Algebra, 2001, 244: 312-342.
9. F. Li, Weak Hopf algebras and some new solutions of the quantum Yang-Baxter equation, J. Algebra, 1998, 208: 72-100.
10. F. Li, Weak Hopf algebras and regular monoids, J. Math. Research and Exposition, 1999, 19: 325-331.
11. F. Li, Solutions of Yang-Baxter equation in endomorphism semigroup and quasi-(co)braided almost bialgebras, Comm. Algebra, 2000, 28: 2253-2270.
12. F. Li, Some weaker structures of Hopf algebras, Journal of Zhejiang University (Science Edition in Chinese), 2002, 29(3).
13. F. Li, On quasi-bicrossed products of weak Hopf algebras, Acta Math. Sinica, 2004, 20(2): 305-318.

14. F. Li, The Right Braids, Quasi-braided pre-tensor categories, and general Yang-Baxter operators, Comm. Algebra, 2004, 32(2): 397-441.

15. F. Li, Structure of Quantum Doubles of Finite Clifford Monoids, Comm. Algebra, 2004, 32(10): 4085-4097.

16. F. Li, H. J. Cao, Semilattice graded weak Hopf algebra and its related quantum G-double. J. Math. Phys., 2005, 46(8): 083519.

17. F. Li, S. Duplij, Weak Hopf algebras and singular solutions of quantum Yang-Baxter equation, Commun. Math. Phys., 2002, 225: 191-217.

18. F. Li, G. X. Liu, Weak tensor category and related generalized Hopf algebras, Acta Math. Sinica, 2006, 22(4): 1027-1046.

19. G. Lusztig, Introduction to Quantum Groups, Progress in Mathematics Birkhauser, Boston, 1993d, Vol. 110.

20. F. Li, Y. Z. Zhang, Quantum doubles from a class of noncocommutative weak Hopf algebras, J. Math. Phys., 2004, 45(8): 3266-3281.

21. S. Majid, Quasitriangular Hopf algebras and Yang-Baxter equations, Internat. J. Modern Phys. A, 1990, 5: 1-91.

22. A. Munir, F. Li, Self-Dual Weak Hopf Algebras, Acta Mathematica Sinica (English Series), 2008, 24(12): 1935-1948.

23. G .Mason, The quantum double of a finite group and its role in conformal field theory, Groups'93, Lecture Note Series 212, London Math. Soc., Cambridge Univ. Press, 1994.

24. F. Nill, Axioms for weak bialgebras, Math. QA/9805104.

25. D. Nikshych, L. Vainerman, A characterization of depth 2 subfactors of II1 factors, J. Func. Analysis, 2000, 171: 278-307.

26. M. E. Sweedler, Hopf Algebras, New York: Benjamin, 1969.

27. M. Takeuchi, Some topics on $GL_q(n)$, J. Algebra, 1992, 147(2): 379-410.

28. Z. X. Wu, The weak Hopf algebras related to generalized Kac-Moody algebra, J. Math. Phys., 2006, 47, 062108.

29. S. H. Wang, T. S. Ma, Singular solutions to the quantum Yang-Baxter equations, Comm. Algebra 2009, 37: 296-316.

30. S. L. Yang, Weak Hopf algebras corresponding to Cartan matrices, J. Math. Phys., 2005, 46: 073502.

31. L. X. Ye, F. Li, Weak quantum Borcherds superalgebras and their representations, J. Math. Phys., 2007, 48: 023502.

32. L. X. Ye, Z. X. Wu, X. F. Mei, Weak Hopf algebras corresponding to Borcherds Cartan Matrices, Acta Mathematica Sinica, 2007, 23(10): 1729-1744.

33. Z. X. Wu, A class of weak Hopf algebras related to a Borcherds-Cartan Matrix, J. Phys. A Math. Gen., 2006, 39: 14611-14626.

34. Z. X. Wu, Weak quantum enveloping algebras of Borcherds superalgebras, Acta Appl. Math., 2009, 106: 185-198.

THE EQUITABLE PRESENTATION FOR THE QUANTUM ALGEBRA $U_q(f(k))$

YAN PAN

*School of Mathematics, Yangzhou University
and Hanjiang Middle School, Yangzhou, 225002, China
E-mail:jspanyan@163.com*

MEILING ZHU* and LIBIN LI[†]

*School of Mathematics, Yangzhou University ,
Yangzhou, 225002, China
*E-mail:147484340@qq.com
† E-mail:lbli@yzu.edu.cn*

We show that the quantum algebra $U_q(f(k))$ has a presentation with generators $x^{\pm 1}, y, z$ and relations

$$xx^{-1} = x^{-1}x = 1,$$

$$\frac{q^m yz - q^{-m}zy}{q^m - q^{-m}} = 1, \quad \frac{xy - q^{-2}yx}{1 - q^{-2}} = x^{1-m}, \quad \frac{xz - q^2 zx}{1 - q^2} = x^{1-m},$$

we call this the *equitable* presentation of $U_q(f(k))$. Then by displaying an infinite dimensional $U_q(f(k))$-module that contains a nonzero null vector for y (resp.z), we obtain that y (resp.z) is not invertible in $U_q(f(k))$. At last, we show that y and z are invertible on each finite dimensional $U_q(f(k))$-module, and satisfy

$$\Omega^{-1}x^m\Omega = y, \quad \Omega^{-1}y\Omega = z, \quad \Omega^{-1}z\Omega = x^m,$$

where Ω is a linear operator defined on finite dimensional module by using some q−exponential function.

Keywords: Quantum group; quantum algebra; equitable presentation.

Introduction

It is well known that the quantized enveloping algebra $U_q(\mathfrak{sl}(2))$ has played a fundamental role in the study of the general quantized enveloping algebras (see [1,5,7,8,9]). In [4] Ito, Weng and Terwilliger introduced the equitable presentation

---

[†]Corresponding author.

for the quantum group $U_q(\mathfrak{sl}(2))$. In [12], Terwilliger gave an analogous presentation for the quantum group $U_q(\mathfrak{g})$ for $\mathfrak{g}$ being a symmetrizable Kac-Moody algebra (see [4,6]). Recently, Smith and many other scholars have introduced a new quantized deformation $U = U_q(f(k))$ ([2, 3, 10,11]). In particular, for some special parameters $f(k) = a(K^m - K^{-m})$, $a \neq 0$, $m \in \mathbb{N}$, the algebra $U_q(f(k))$ is quantum group, all the finite dimensional representations of $U_q(f(k))$ are proven to be semi-simple and the irreducible $U_q(f(k))$-module is also constructed in [2, 12]. Based on [4, 12], we give the equitable generators $x^{\pm 1}, y, z$ for the quantum group $U_q(f(k))$ and show that y, z are not invertible in $U_q(f(k))$. At last, we prove that y and z are invertible on each finite dimensional $U_q(f(k))$-module, and satisfy

$$\Omega^{-1} x^m \Omega = y, \quad \Omega^{-1} y \Omega = z, \quad \Omega^{-1} z \Omega = x^m,$$

where Ω is a linear operator defined on finite dimensional module by using some q-exponential function.

1. The Algebra $U_q(f(k))$

Let $\mathbb{C}$ be the field of complex numbers and q be an indeterminate. We will work over the field $\mathbb{C}(q)$. Let m be a fixed integer, and α be a $2m$-th root of unity. For an integer n we define:

$$[n] = \frac{q^n - q^{-n}}{q - q^{-1}}, \qquad [n]_m = \frac{q^{nm} - q^{-nm}}{q^m - q^{-m}},$$

and for $n \geq 0$ we define:

$$[n]! = [n][n-1]\cdots[2][1], \qquad [n]_m! = [n]_m[n-1]_m\cdots[2]_m[1]_m,$$

$$\begin{bmatrix} n \\ i \end{bmatrix} = \frac{[n]!}{[i]![n-i]!}, \qquad \begin{bmatrix} n \\ i \end{bmatrix}_m = \frac{[n]_m!}{[i]_m![n-i]_m!}.$$

We interpret $[0]! = 1$. Recall that the quantum algebra $U_q(f(k))$ is defined as follows:

Definition 1.1. *For any Laurent polynomial* $f(k) = a(K^m - K^{-m}) \in \mathbb{C}(q)[k, k^{-1}]$, $a \neq 0$, $m \in \mathbb{N}$, *let* $U_q(f(k))$ *denote the associative* $\mathbb{C}(q)$-*algebra with generators* k, k^{-1}, e, f *and the following relations:*

$$kk^{-1} = k^{-1}k = 1, \quad ke = q^2 ek, \quad kf = q^{-2} fk, \quad ef - fe = f(k).$$

We call $k^{\pm 1}, e, f$ the *Chevalley generators* for $U_q(f(k))$. In the following, we always assume that $f(k) = \frac{k^m - k^{-m}}{q - q^{-1}}$ for $0 < m \in \mathbb{N}$. It is well known that the quantum group $U_q(f(k))$ has a unique Hopf algebra structure, the comultiplication Δ is given by

$$\Delta(k) = k \otimes k, \ \Delta(e) = e \otimes k^m + 1 \otimes e, \ \Delta(f) = f \otimes 1 + k^{-m} \otimes f;$$

the counit ε is given by

$$\varepsilon(k) = 1, \quad \varepsilon(e) = \varepsilon(f) = 0.$$

and the antipode s is given by

$$s(k) = k^{-1}, \quad s(e) = -ek^{-m}, \quad s(f) = -k^m f.$$

We refer the reader to [2,3,11] for background information on $U_q(f(k))$.

2. The Equitable Presentation for $U_q(f(k))$

In the presentation for $U_q(f(k))$ given in Definition 1.1 the generators $k^{\pm 1}$ and the generators e, f play a very different role. In this section, we introduce a presentation for $U_q(f(k))$ whose generators are on a more equal footing.

Theorem 2.1. *The algebra $U_q(f(k))$ is generated by $x^{\pm 1}, y, z$, where*

$$y = k^{-m} + f(q^m - q^{-m}),$$
$$z = k^{-m} - k^{-m} e q^m (q - q^{-1}),$$
$$x^{\pm 1} = k^{\pm 1},$$

and the generators $x^{\pm 1}, y, z$ satisfy the following relations:

$$xx^{-1} = x^{-1}x = 1, \tag{1}$$

$$\frac{q^m yz - q^{-m} zy}{q^m - q^{-m}} = 1, \tag{2}$$

$$\frac{xy - q^{-2} yx}{1 - q^{-2}} = x^{1-m}, \tag{3}$$

$$\frac{xz - q^2 zx}{1 - q^2} = x^{1-m}. \tag{4}$$

Proof. One readily checks that the following relations in $U_q(f(k))$ hold:

$$e = (1 - x^m z) q^{-m} (q - q^{-1})^{-1},$$
$$f = (y - x^{-m})(q^m - q^{-m})^{-1},$$
$$k^{\pm 1} = x^{\pm 1}.$$

The assertions follow using the definition relations in section 2. □

We see that the generators $x^{\pm 1}, y, z$ from Theorem 2.1 are on an equal footing, more or less, in view of this we make a definition as follows:

Definition 2.1. *By the equitable presentation for $U_q(f(k))$ we mean the presentation given in Theorem 2.1. We call $x^{\pm 1}, y, z$ the equitable generators.*

3. The Elements y and z Are Not Invertible in $U_q(f(k))$

In this section we show that the equitable generators y and z are not invertible in $U_q(f(k))$. In order to show that y(resp.z) is not invertible in $U_q(f(k))$ we display an infinite dimensional $U_q(f(k))$-module that contains a nonzero null vector for y (resp.z).

Lemma 3.1. *There exists a $U_q(f(k))$-module Γ_y with the following property: Γ_y has a basis*

$$u_{ij} \quad i,j \in \mathbb{Z}, \ j \geq 0$$

such that

$$
\begin{aligned}
xu_{ij} &= u_{i+1,j}, \\
x^{-1}u_{ij} &= u_{i-1,j}, \\
yu_{ij} &= (q^{2i} - q^{2i-2j})u_{i,j-m} - (q^{2i} - 1)u_{i-m,j}, \\
zu_{ij} &= q^{-2i}u_{i,j+m} + (1 - q^{-2i})u_{i-m,j}.
\end{aligned}
$$

for all $i,j \in \mathbb{Z}$ with $j \geq 0$, where $u_{i,j} := 0$ for $j < 0$.

Proof. One routinely verifies that the given actions of $x^{\pm 1}, y, z$ satisfy the relations $(1) - (4)$. $\qquad \square$

Lemma 3.2. *The following $(i) - (iii)$ hold:*

(i) $yu_{00} = 0$, where the vector u_{00} is from Lemma 3.1;
(ii) y is not invertible on Γ_y, where Γ_y is the $U_q(f(k))$-module from Lemma 3.1;
(iii) y is not invertible in $U_q(f(k))$.

Proof. It is immediate from Lemma 3.1 above. $\qquad \square$

Remark: Referring to Lemma 3.1, we have $u_{ij} = x^i z^b u_{0a}$ for $j = a + bm, 0 \leq a < m$, and $b > 0$.

Lemma 3.3. *There exists a $U_q(f(k))$-module Γ_z with the following property: Γ_z has a basis*

$$v_{ij} \quad i,j \in \mathbb{Z}, j \geq 0$$

such that

$$
\begin{aligned}
xv_{ij} &= v_{i+1,j}, \\
x^{-1}v_{ij} &= v_{i-1,j}, \\
yv_{ij} &= q^{2i}v_{i,j+m} - (q^{2i} - 1)v_{i-m,j}, \\
zv_{ij} &= (1 - q^{-2i})v_{i-m,j} - (q^{2j-2i} - q^{-2i})v_{i,j-m}.
\end{aligned}
$$

for all $i,j \in \mathbb{Z}$ with $j \geq 0$, where $v_{i,j} := 0$ for $j < 0$.

Proof. One routinely verifies that the given actions of $x^{\pm 1}, y, z$ satisfy the relations $(1) - (4)$. $\square$

Lemma 3.4. *The following $(i) - (iii)$ hold.*

(i) $zv_{00} = 0$, *where the vector* v_{00} *is from Lemma 3.3;*
(ii) z *is not invertible on* Γ_z, *where* Γ_z *is the* $U_q(f(k))$-*module from Lemma 3.3;*
(iii) z *is not invertible in* $U_q(f(k))$.

Proof. Easy. $\square$

Remark: Referring to Lemma 3.3, we have $v_{ij} = x^i y^b v_{0a}$ for $j = a + bm, 0 \le a < m, b > 0$.

4. The Elements y and z Are Invertible on Each Finite Dimensional $U_q(f(k))$-Module

In this section we show that the equitable generators y and z are invertible on each finite dimensional $U_q(f(k))$-module. We begin with some general comments. By [2] each finite dimensional $U_q(f(k))$-module M is semi-simple, that is, M is a direct sum of simple $U_q(f(k))$-modules. The list of finite dimensional simple $U_q(f(k))$-modules is described as follows.

Lemma 4.1. *For an integer* $n \ge 0$, *there exists a finite dimensional simple* $U_q(f(k))$-*module* $V(n, \alpha)$ *with the following properties:* $V(n, \alpha)$ *has a basis* $v_0, v_1, \cdots, v_n$ *such that*

$$kv_i = \alpha q^{n-2i} v_i,$$
$$ev_i = \alpha^m [n - i + 1]_m [m] v_{i-1},$$
$$fv_i = [i + 1]_m v_{i+1}, \quad 0 \le i \le n,$$

where $ev_{-1} = fv_{n+1} = 0$. *Moreover, every finite dimensional simple* $U_q(f(k))$-*module with dimension* $n + 1$ *is isomorphic to* $V(n, \alpha)$ *for some* α *being* $2m-$ *th root of unity* .

The equitable generators act on the module $V(n, \alpha)$ as follows.

Lemma 4.2. *For an integer* $n \ge 0$, *the* $U_q(f(k))$-*module* $V(n, \alpha)$ *has a basis* $u_0, u_1, \cdots, u_n$ *such that*

$$xu_i = \alpha q^{n-2i} u_i \quad (0 \le i \le n), \tag{5}$$
$$x^m u_i = \alpha^m q^{(n-2i)m} u_i \quad (0 \le i \le n), \tag{6}$$
$$(\alpha^{-m} y - q^{(2i-n)m}) u_i = (q^{-nm} - q^{(2i+2-n)m}) u_{i+1} \quad (0 \le i \le n), \tag{7}$$
$$(\alpha^{-m} z - q^{(2i-n)m}) u_i = (q^{nm} - q^{(2i-2-n)m}) u_{i-1} \quad (0 \le i \le n). \tag{8}$$

where we set $u_{-1} = u_{n+1} = 0$.

Proof. Let $v_0, v_1, \cdots, v_n$ be the basis for $V(n, \alpha)$ as in Lemma 4.1. Define $u_i = \gamma_i v_i$ for $0 \le i \le n$, where $\gamma_0 = 1$ and $\gamma_i = -\alpha^m q^{(n-i)m} \gamma_{i-1}$ for $1 \le i \le n$. Using $x = k, y = k^{-m} + f(q^m - q^{-m})$ and $z = k^{-m} - k^{-m} e q^m (q - q^{-1})$, together with the data in Lemma 4.1, we can routinely verity $(5) - (8)$. $\qquad \square$

Remark: The basis $u_0, u_1, \cdots, u_n$ in Lemma 4.2 is normalized so that $yu = \alpha^m q^{-nm} u$ and $zu = \alpha^m q^{nm} u$ for $u = \sum_{i=0}^{n} u_i$.

Theorem 4.1. *For an integer $n \ge 0$, the following $(i), (ii)$ hold on the $U_q(f(k))$-module $V(n, \alpha)$.*

  (i) Each of x^m, y, z is semi-simple with eigenvalues

$$\alpha^m q^{nm}, \alpha^m q^{(n-2)m}, \cdots, \alpha^m q^{-nm}.$$

  (ii) The actions of x, y, z are invertible.

Proof. For x, this is clear from (5). We now verify our assertions for y. With respect to the basis $u_0, u_1, \cdots, u_n$ for $V(n, \alpha)$ given in Lemma 4.2, by (7) we obtain that

$$\alpha^{-m} y u_i = q^{(2i-n)m} u_i + (q^{-nm} - q^{(2i+2-n)m}) u_{i+1},$$

then we have the matrix representing y :

$$
y \begin{pmatrix} u_0 \\ u_1 \\ \vdots \\ u_n \end{pmatrix} = \alpha^m \begin{pmatrix} q^{-nm} & q^{-nm} - q^{(2-n)m} & \cdots & 0 \\ 0 & q^{(2-n)m} & \cdots & 0 \\ \vdots & \vdots & \ddots & \vdots \\ 0 & 0 & \vdots & q^{nm} \end{pmatrix} \begin{pmatrix} u_0 \\ u_1 \\ \vdots \\ u_n \end{pmatrix}
$$

It is clear that the matrix representing y is lower triangular with (i, i) entry $\alpha^m q^{(2i-n)m}$ for $0 \le i \le n$. Therefore the action of y on $V(n, \alpha)$ has eigenvalues $\alpha^m q^{nm}, \alpha^m q^{(n-2)m}, \cdots, \alpha^m q^{-nm}$. These eigenvalues are mutually distinct so this action is semi-simple. These eigenvalues are nonzero so the action of y is invertible. It follows that the assertions hold for y. the assertions for z are similarly verified. $\qquad \square$

By Theorem 4.1 and the fact that each finite dimensional $U_q(f(k))$-module is semi-simple we obtain the following result.

Corollary 4.1. *On each finite dimensional $U_q(f(k))$-module the actions of y and z are invertible. Let y^{-1} (resp. z^{-1}) denote the linear operator that acts on each finite dimensional $U_q(f(k))$-module as the inverse of y (resp. z).*

5. The Elements n_x, n_y, n_z

In this section we define some elements n_x, n_y, n_z in $U_q(f(k))$ and show that these are nilpotent on each finite dimensional $U_q(f(k))$-module. We then recall the q-exponential function exp_{q^m} and derive a number of equations involving $exp_{q^m}(n_x)$, $exp_{q^m}(n_y), exp_{q^m}(n_z)$. Using these equations we will show that on finite dimensional $U_q(f(k))$-modules the operators y^{-1}, z^{-1} from Corollary 4.1 satisfy

$$y^{-1} = exp_{q^m}(n_z)x^m exp_{q^m}(n_z)^{-1} \tag{9}$$

$$z^{-1} = exp_{q^m}(n_y)^{-1}x^m exp_{q^m}(n_y) \tag{10}$$

We begin with an observation.

Definition 5.1. *Let n_x, n_y, n_z be the following elements in $U_q(f(k))$:*

$$n_x = \frac{q^m(1-yz)}{q^m - q^{-m}} = \frac{q^{-m}(1-zy)}{q^m - q^{-m}}, \tag{11}$$

$$n_y = \frac{q^m(1-zx^m)}{q^m - q^{-m}} = \frac{q^{-m}(1-x^mz)}{q^m - q^{-m}}, \tag{12}$$

$$n_z = \frac{q^m(1-x^my)}{q^m - q^{-m}} = \frac{q^{-m}(1-yx^m)}{q^m - q^{-m}}. \tag{13}$$

We recall some notations. Let V denote a finite dimensional vector space over $\mathbb{C}$. A linear transformation $T : V \to V$ is called nilpotent whenever there exists a positive integer r such that $T^r V = 0$. We are going to show that each of n_x, n_y, n_z is nilpotent on all finite dimensional $U_q(f(k))$- modules. We will show this fact using the following lemma.

Lemma 5.1. *The following relations hold in $U_q(f(k))$:*

$$xn_y = q^{2m}n_yx, \quad x^mn_z = q^{-2m}n_zx^m, \tag{14}$$

$$yn_z = q^{2m}n_zy, \quad yn_x = q^{-2m}n_xy, \tag{15}$$

$$zn_x = q^{2m}n_xz, \quad zn_y = q^{-2m}n_yz. \tag{16}$$

Proof. In order to verify these equations, eliminate n_x, n_y, n_z using Definition 5.1 and simplify the result. □

Theorem 5.1. *Each of n_x, n_y, n_z is nilpotent on all finite dimensional $U_q(f(k))$-modules.*

Proof. We prove the result for n_x, the proof for n_y and n_z is similar. Since each finite dimensional $U_q(f(k))$-module is semi-simple and in view of Lemma 4.1, it suffices to show that n_x is nilpotent on each module $V(n, \alpha)$. By Theorem 4.1, $V(n, \alpha)$ has a basis $w_0, w_1, \cdots, w_n$ such that

$$yw_i = \alpha^m q^{(n-2i)m} w_i$$

for $0 \le i \le n$. Using the equation on the right in (12) we routinely find that

$$yn_x w_i = q^{-2m} n_x y w_i = q^{-2m} n_x \alpha^m q^{(n-2i)m} w_i = \alpha^m q^{(n-2(i+1))m} n_x w_i,$$

that is, $n_x w_i$ is a scalar multiple of w_{i+1} for $0 \le i \le n-1$ and $n_x w_n = 0$. This shows that n_x is nilpotent on $V(n, \alpha)$ and the result follows. $\qquad\square$

Definition 5.2. *Let T denote a linear operator that acts on all finite dimensional $U_q(f(k))$-modules in a nilpotent fashion. We define:*

$$exp_{q^m}(T) = \Sigma_{i=0}^{\infty} \frac{q^{i(i-1)m/2}}{[i]_m!} T^i. \tag{17}$$

We view $exp_{q^m}(T)$ as a linear operator that acts on all finite dimensional $U_q(f(k))$-modules. It is clear that $exp_{q^m}(T)$ is invertible on each finite dimensional $U_q(f(k))$-module, and the inverse is:

$$exp_{q^{-m}}(-T) = \Sigma_{i=0}^{\infty} \frac{(-1)^i q^{-i(i-1)m/2}}{[i]_m!} T^i.$$

Lemma 5.2. *The following $(i) - (iii)$ hold on each finite dimensional $U_q(f(k))$-module.*

(i) $exp_{q^m}(n_y)^{-1} x^m exp_{q^m}(n_y) = z^{-1}$,

(ii) $exp_{q^m}(n_z)^{-1} y exp_{q^m}(n_z) = x^{-m}$,

(iii) $exp_{q^m}(n_x)^{-1} z exp_{q^m}(n_x) = y^{-1}$.

Proof. (i) We show

$$x^m exp_{q^m}(n_y) z = exp_{q^m}(n_y). \tag{18}$$

The left side of (18) is equal to $x^m exp_{q^m}(n_y) x^{-m} x^m z$. Observe that

$$x^m exp_{q^m}(n_y) x^{-m} = exp_{q^m}(x^m n_y x^{-m})$$

by (18) and $x^m n_y x^{-m} = q^{2m} n_y$ by (14). Also $x^m z = 1 - q^m(q^m - q^{-m})n_y$ by (13). Using these comments and (17) we routinely find that the left side of (18) is equal to the right side of (18). The result follows. It is similar to prove (ii) and (iii). $\qquad\square$

For convenience we display a second version of Lemma 5.2.

Lemma 5.3. *The following (i) – (iii) hold on each finite dimensional $U_q(f(k))$-module.*

(i) $exp_{q^m}(n_z)x^m exp_{q^m}(n_z)^{-1} = y^{-1}$,

(ii) $exp_{q^m}(n_x)yexp_{q^m}(n_x)^{-1} = z^{-1}$,

(iii) $exp_{q^m}(n_y)zexp_{q^m}(n_y)^{-1} = x^{-m}$.

Proof. For each of the equations in Lemma 5.2 take the inverse of each side and simplify the result. □

We know that the equations (9), (10) are just Lemma 5.3(i) and Lemma 5.2(i), respectively.

6. Some Formulae Involving the q-Exponential Function

In the next section we will display a linear operator Ω that acts on all finite dimensional $U_q(f(k))$- modules, and satisfies

$$\Omega^{-1}x^m\Omega = y, \ \Omega^{-1}y\Omega = z, \ \Omega^{-1}z\Omega = x^m$$

on these modules. In order to prove the desired properties we will first establish a few identities. These identities are given in this section.

Lemma 6.1. *The following (i) – (iii) hold on each finite dimensional $U_q(f(k))$-module.*

(i) $exp_{q^m}(n_z)^{-1}x^m exp_{q^m}(n_z) = x^m yx^m$,

(ii) $exp_{q^m}(n_x)^{-1}yexp_{q^m}(n_x) = yzy$,

(iii) $exp_{q^m}(n_y)^{-1}zexp_{q^m}(n_y) = zx^m z$.

Proof. (*i*) The element $x^m y$ commutes with n_z by (13) so $x^m y$ commutes with $exp_{q^m}(n_z)$ in view of (18). Therefore $exp_{q^m}(n_z)^{-1}x^m yexp_{q^m}(n_z) = x^m y$. By Lemma 5.3(i) we have $yexp_{q^m}(n_z) = exp_{q^m}(n_z)x^{-m}$. Combining these last two equations we routinely obtain the result. It is similar to prove (*ii*) and (*iii*). □

Lemma 6.2. *The following (i) – (iii) hold on each finite dimensional $U_q(f(k))$-module.*

(i) $exp_{q^m}(n_y)x^m exp_{q^m}(n_y)^{-1} = x^m z x^m,$

(ii) $exp_{q^m}(n_z)y exp_{q^m}(n_z)^{-1} = y x^m y,$

(iii) $exp_{q^m}(n_x)z exp_{q^m}(n_x)^{-1} = z y z.$

Proof. (i) By the Lemma 6.1(iii) we have $exp_{q^m}(n_y)zx^m z exp_{q^m}(n_y)^{-1} = z$. In this equation we eliminate $exp_{q^m}(n_y)z$ and $z exp_{q^m}(n_y)^{-1}$ using Lemma 5.3(iii). The result follows. (ii), (iii) Similar to the proof of (i) above. $\square$

Lemma 6.3. *The following* (i) – (iii) *hold on each finite dimensional* $U_q(f(k))$-*module.*

(i) $exp_{q^m}(n_x)^{-1}x^m exp_{q^m}(n_x) = x^m + y - y^{-1},$

(ii) $exp_{q^m}(n_y)^{-1}y exp_{q^m}(n_y) = y + z - z^{-1},$

(iii) $exp_{q^m}(n_z)^{-1}z exp_{q^m}(n_z) = z + x^m - x^{-m}.$

Proof. (i) Using (2), (4) and (11) we obtain $x^m n_x - n_x x^m = y - z$. By this and a routine induction using (15), (16) we find

$$x^m n_x^i - n_x^i x^m = q^{(1-i)m}[i]_m (n_x^{i-1}y - zn_x^i). \tag{19}$$

for each integer $i \geq 0$. Using (17) and (19) we obtain

$$x^m exp_{q^m}(n_x) - exp_{q^m}(n_x)x^m = exp_{q^m}(n_x)y - z exp_{q^m}(n_x). \tag{20}$$

In line (20) we multiply each term on the left by $exp_{q^m}(n_x)^{-1}$ and evaluate the term containing z using Lemma 5.2(iii) to get the result. It is similar to prove (ii) and (iii). $\square$

Lemma 6.4. *The following* (i) – (iii) *hold on each finite dimensional* $U_q(f(k))$-*module.*

(i) $exp_{q^m}(n_x)x^m exp_{q^m}(n_x)^{-1} = x^m + z - z^{-1},$

(ii) $exp_{q^m}(n_y)y exp_{q^m}(n_y)^{-1} = y + x^m - x^{-m},$

(iii) $exp_{q^m}(n_z)z exp_{q^m}(n_z)^{-1} = z + y - y^{-1}.$

Proof. (i) By Lemma 6.3(i) we have

$$x^m = exp_{q^m}(n_x)(x^m + y - y^{-1})exp_{q^m}(n_x)^{-1}. \tag{21}$$

By Lemma 5.3(ii) we have $exp_{q^m}(n_x)yexp_{q^m}(n_x)^{-1} = z^{-1}$ and

$$exp_{q^m}(n_x)y^{-1}exp_{q^m}(n_x)^{-1} = z.$$

Evaluating (21) using these comments we obtain the result. (ii), (iii) Similar to the proof of (i) above. $\qquad\square$

7. The Operator Ω

In this section we display a linear operator Ω that acts on all finite dimensional $U_q(f(k))$-modules, and satisfies

$$\Omega^{-1}x^m\Omega = y, \quad \Omega^{-1}y\Omega = z, \quad \Omega^{-1}z\Omega = x^m$$

on these modules. In order to define Ω we first recall the notion of a *weight space*.

Definition 7.1. *Let M be a finite dimensional $U_q(f(k))$-module. For an integer λ, define*

$$M(\alpha^m, \lambda) = \{v \in M | x^m v = \alpha^m q^{\lambda m} v\}.$$

We call $M(\alpha^m, \lambda)$ the (α^m, λ)-weight space of M with respect to x^m. By Theorem 4.1(i) and since M is semi-simple, M is the direct sum of its weight spaces with respect to x^m.

Define a linear operator ψ that acts on each finite dimensional $U_q(f(k))$-module M. In order to do this we give the action of ψ on each weight space of M with respect to x^m. For an integer λ, ψ acts on the weight space $M(\alpha^m, \lambda)$ as $q^{-\lambda^2 m/2}I$ (if λ is even) and $q^{(1-\lambda^2)m/2}I$ (if λ is odd), where I denotes the identity map. We observe that ψ is invertible on M.

Lemma 7.1. *For the operator ψ, the following $(i) - (iii)$ hold on each finite dimensional $U_q(f(k))$-module.*

(i) $\psi^{-1}x\psi = x$,

(ii) $\psi^{-1}n_y\psi = x^m n_y x^m$,

(iii) $\psi^{-1}n_z\psi = x^{-m}n_z x^{-m}$

Proof. Let $\dot{M}$ denote a finite dimensional $U_q(f(k))$-module. For an integer λ, we show that each of $(i) - (iii)$ holds on $M(\alpha^m, \lambda)$.

(i) On $M(\alpha^m, \lambda)$ each of ψ, x acts as a scalar multiple of the identity.

(ii) For notational convenience define $s = 0$ (if λ is even) and $s = 1$ (if λ is odd). For $v \in M(\alpha^m, \lambda)$ we show $\psi^{-1} n_y \psi v = x^m n_y x^m v$. Using the equation on the left in (14) we find $n_y v \in M(\alpha^m, \lambda + 2)$. Using this we find

$$\psi^{-1} n_y \psi v = q^{(s-\lambda^2)m/2} \psi^{-1} n_y v = q^{(2\lambda+2)m} n_y v$$

and also

$$x^m n_y x^m v = \alpha^m x^m n_y q^{\lambda m} v = \alpha^m q^{\lambda m} x^m (n_y v) = \alpha^m q^{\lambda m} \alpha^m q^{(\lambda+2)m} n_y v = q^{(2\lambda+2)m} n_y v.$$

Therefore $\psi^{-1} n_y \psi v = x^m n_y x^m v$. We have now shown $\psi^{-1} n_y \psi$ and $x^m n_y x^m$ coincide on $M(\alpha^m, \lambda)$.

(iii) Similar to the proof of (ii) above. □

Definition 7.2. *We define*

$$\Omega = exp_{q^m}(n_z) \psi exp_{q^m}(n_y) \tag{22}$$

where n_y, n_z are from Definition 5.1. We view Ω as a linear operator that acts on all finite dimensional $U_q(f(k))$-module.

We now present our main result.

Theorem 7.1. *For the operator Ω from Definition 7.2 the following hold on each finite dimensional $U_q(f(k))$-module:*

$$\Omega^{-1} x^m \Omega = y, \quad \Omega^{-1} y \Omega = z, \quad \Omega^{-1} z \Omega = x^m.$$

Proof. We only show the first identity, the others can be proven similarly.

$$\begin{aligned}
\Omega^{-1} x^m \Omega &= exp_{q^m}(n_y)^{-1} \psi^{-1} exp_{q^m}(n_z)^{-1} x^m exp_{q^m}(n_z) \psi exp_{q^m}(n_y) \\
&= exp_{q^m}(n_y)^{-1} \psi^{-1} x^m y x^m \psi exp_{q^m}(n_y) \\
&= exp_{q^m}(n_y)^{-1} \psi^{-1} x^m y \psi x^m exp_{q^m}(n_y) \\
&= exp_{q^m}(n_y)^{-1} \psi^{-1} (1 - q^{-m}(q^m - q^{-m})n_z) \psi x^m exp_{q^m}(n_y) \\
&= exp_{q^m}(n_y)^{-1} (x^m - q^{-m}(q^m - q^{-m})x^{-m} n_z) exp_{q^m}(n_y) \\
&= exp_{q^m}(n_y)^{-1} (x^m - x^{-m} + y) exp_{q^m}(n_y) \\
&= y.
\end{aligned}$$

□

Corollary 7.1. *On a finite dimensional $U_q(f(k))$-module, we have the following equations*

$$\Omega^3 x^m = x^m \Omega^3, \quad \Omega^3 y = y \Omega^3, \quad \Omega^3 z = z \Omega^3.$$

Proof. By the Theorem 7.1 we know $\Omega x^m = z\Omega, \Omega y = x^m\Omega, \Omega z = y\Omega$, then we have

$$\Omega^3 x^m = \Omega^2(\Omega x^m) = \Omega^2(z\Omega) = \Omega(\Omega z)\Omega = \Omega(y\Omega)\Omega = (\Omega y)\Omega^2 = x^m\Omega^3.$$

Similarly, we can show that $\Omega^3 y = y\Omega^3, \quad \Omega^3 z = z\Omega^3.$ □

In the following, we will describe the action of Ω on the module $V(n, \alpha)$. We will do this by displaying the actions of Ω and Ω^{-1} on the basis for $V(n, \alpha)$ given in Lemma 4.2. We begin with a few observations.

Lemma 7.2. *For an integer $n \geq 0$ let $u_0, u_1, \cdots, u_n$ denote the basis for $V(n, \alpha)$ given in Lemma 4.2. Then $\psi u_i = q^{(2i(n-i)+(s-n^2)/2)m}u_i$ for $0 \leq i \leq n$, where $s = 0$ (if n is even) and $s = 1$ (if n is odd).*

Proof. Immediate from the definition of ψ. □

Lemma 7.3. *For an integer $n \geq 0$, let $u_0, u_1, \cdots, u_n$ denote the basis for $V(n, \alpha)$ given in Lemma 4.2. Then the following $(i), (ii)$ hold.*

(i) $n_y u_i = -q^{(n-i)m}[n - i + 1]_m u_{i-1}(1 \leq i \leq n), n_y u_0 = 0,$

(ii) $n_z u_i = q^{-mi}[i + 1]_m u_{i+1}(0 \leq i \leq n - 1), n_z u_n = 0.$

Proof. Immediate from Lemma 4.2 and Definition 5.1. □

Lemma 7.4. *For an integer $n \geq 0$ let $u_0, u_1, \cdots, u_n$ denote the basis for $V(n, \alpha)$ given in Lemma 4.2. Then for $0 \leq j \leq n$ we have*

$$exp_{q^m}(n_y)u_j = \Sigma_{i=0}^j (-1)^{i+j} q^{(n-i-1)(j-i)m} \begin{bmatrix} n - i \\ j - i \end{bmatrix}_m u_i, \tag{23}$$

$$exp_{q^m}(n_y)^{-1} u_j = \Sigma_{i=0}^j q^{(n-j)(j-i)m} \begin{bmatrix} n - i \\ j - i \end{bmatrix}_m u_i, \tag{24}$$

$$exp_{q^m}(n_z)u_j = \Sigma_{i=j}^n q^{j(j-i)m} \begin{bmatrix} i \\ j \end{bmatrix}_m u_i, \tag{25}$$

$$exp_{q^m}(n_z)^{-1} u_j = \Sigma_{i=j}^n (-1)^{i+j} q^{(i-1)(j-i)m} \begin{bmatrix} i \\ j \end{bmatrix}_m u_i \tag{26}$$

Proof. In order to verify (23) and (25), evaluate the left-hand side using Lemma 7.3 and Definition 5.1. Lines (24) and (26) are similarly verified using Lemma 5.3. □

Theorem 7.2. *For an integer $n \geq 0$ let $u_0, u_1, \cdots, u_n$ denote the basis for $V(n, \alpha)$ given in Lemma 4.2. Then for $0 \leq j \leq n$ we have*

$$\Omega u_j = \Sigma_{i=0}^{n-j}(-1)^j q^{((n-i-1)j+(s-n^2)/2)m} \begin{bmatrix} n-i \\ j \end{bmatrix}_m u_i, \qquad (27)$$

$$\Omega^{-1} u_j = \Sigma_{i=n-j}^{n}(-1)^{n-j} q^{((1-i)(n-j)+(n^2-s)/2)m} \begin{bmatrix} i \\ n-j \end{bmatrix}_m u_i. \qquad (28)$$

Proof. In order to verify (27), evaluate the left-hand side using (22), Lemma 7.2, (23), (25), and simplify the result using the Theorem 8.4 in [4]. Line (28) is similar to be verified. □

Acknowledgments

The work was supported by the National Natural Science Foundation of China (Grant No.10771182) and Doctorate Foundation Ministry of Education of China (Grant No. 200811170001).

References

1. C. Kassel, Quantum Groups, Graduate Texts in Mathematics 155, Springer-Verlag, New York, 1995.
2. Q. Z. Ji and D. G. Wang. Finite-Dimensional Representations of Quantum Groups $U_q(f(k))$, East-West J. Math. 2000, 2(2), 201-213.
3. N. H. Jing and J. Zhang, Quantum Wely algebras and deformations of $U(G)$, Pacific J. Math. 1995, 171(2): 437-454.
4. T. Ito, P. Terwilliger and C.W. Weng, The quantum algebra $U_q(sl(2))$ and its equitable presentation, J. Algebra, 2006, 298(1): 284-301.
5. J. C. Jantzen, Lections on quantum groups. Graduate Students in Mathematics 6, Amer. Math. Soc., Providence RI, 1996.
6. V. Kac, Infinite dimensional Lie algebras, Third Edition, Cambridge University Press, Cambridge, 1990.
7. G. Lusztig, Quantum deformation of certain simple modules over enveloping algebras, Adv. Math. 1988, 70: 237-249.
8. G. Lusztig, Introduction to quantum groups. Progress in Mathematics, 110, Birkhäuser, Boston, 1993.
9. S. Montgomery, Hopf algebras and their actions on rings, CBMS series in Math., 1993.
10. S. P. Smith, A class of algebras similar to the enveloping algebra of $sl(2)$, Trans. Amer. Math. Soc. 1990, 322: 285-314.
11. X. Tang, On irreducible weight representations of a new deformation $U_q(sl_2)$ of $U(sl_2)$, J. Gen. Lie Theory Appl. 2007, 1(2), 97–105.
12. P. Terwilliger, The equitable presentation for the quantum group $U_q(g)$ associated with a symmetrizable Kac-Moody algebra g, J. Algebra, 2006, 298(1): 302-319.